John Hunter

The Natural History of the Human Teeth

Explaining their structure, use, formation, growth, and diseases

John Hunter

The Natural History of the Human Teeth
Explaining their structure, use, formation, growth, and diseases

ISBN/EAN: 9783337367978

Printed in Europe, USA, Canada, Australia, Japan

Cover: Foto ©berggeist007 / pixelio.de

More available books at **www.hansebooks.com**

THE

NATURAL HISTORY

OF

THE HUMAN TEETH.

EXPLAINING THEIR STRUCTURE, USE, FORMATION,

GROWTH, AND DISEASES.

IN TWO PARTS.

By JOHN HUNTER, F.R.S.

WITH NOTES BY

FRANCIS C. WEBB, M.D., F.L.S.,

MEMBER OF THE ROYAL COLLEGE OF PHYSICIANS, LONDON, PHYSICIAN TO THE GREAT NORTHERN HOSPITAL, ETC.;

AND

ROBERT T. HULME, M.R.C.S., F.L.S.,

LECTURER ON DENTAL SURGERY AT THE METROPOLITAN SCHOOL OF DENTAL SCIENCE, AND DENTAL SURGEON TO THE NATIONAL DENTAL HOSPITAL.

LONDON:
ROBERT HARDWICKE, 192 PICCADILLY.

MDCCCLXV.

NOTICE.

THE present edition of HUNTER ON THE TEETH originally appeared in the pages of the "Dental Review" for the years 1861—63.

The Notes appended to the First Part were furnished by Dr. WEBB, and formed the basis of a Course of Lectures on the Comparative Anatomy of the Teeth, delivered to the Students of the Metropolitan School of Dental Science. Such notes were necessarily a work of compilation rather than of original research, and have been principally derived from the rich stores of information to be found in the works of OWEN, TOMES, NASMYTH, and KÖLLIKER. Obligations to these and other authors are duly acknowledged.

The Notes to the Second Part were furnished by Mr. HULME, and are restricted to short comments on Hunter's views. To have attempted more, and to have brought this portion of the work up to the present advanced state of Dental Surgery, would have necessitated the writing an entire treatise on the subject.

In the present edition, the Annotators' Notes are distinguished from those of the Author by alphabetical references and by being placed within brackets.

CONTENTS.

PART I.

	PAGE
Of the Upper Jaw	1
Lower Jaw	3
Alveolar Processes	7
Articulation of the Lower Jaw	9
Motion in the Joint of the Lower Jaw	13
Muscles of the Lower Jaw	16
Temporals	19
Internal Pterygoid Muscle	22
External Pterygoid Muscle	22
Digastric Muscle	26
Structure of a Tooth	34
Enamel	34
Bony Part of a Tooth	44
Cavity of the Teeth	53
Periosteum of the Teeth	55
Situation of the Teeth	56
Number of Teeth	57
Incisores	61
Cuspidatus	70
Bicuspides	78
Grinders	81
Articulation of the Teeth	105
Gums	106
Action of the Teeth arising from the Motion of the Lower Jaw	107
General Comparisons between the Motion of the Jaw in Young and in Old People	109
Of the Formation of the Alveolar Process	110
Formation of the Teeth in the Fœtus	115
Cause of Pain in Dentition	120
Formation and Progress of the Adult Teeth	121
Manner in which a Tooth is formed	127
Ossification of a Tooth upon the Pulp	129
Formation of the Enamel	135
Manner of Shedding the Teeth	140

	PAGE
Of the Growth of the Two Jaws	141
The Reason for the Shedding of the Teeth	143
Of the Cavity filling up as the Teeth wear down	145
Continual Growth of the Teeth	146
Sensibility of the Teeth	149
Supernumerary Teeth	149
The Use of the Teeth so far as they affect the Voice	151
Of the Class among Animals of the Human Teeth	152
Diseases of the Teeth	153
Cleaning the Teeth	154
Transplanting the Teeth	156

PART II.

INTRODUCTION.

CHAP. I.—OF THE DISEASES OF THE TEETH AND THE CONSEQUENCES OF THEM 161

1. The Decay of the Teeth arising from Rottenness—Symptoms of Inflammation—Stopping of the Teeth.—2. The Decay of the Teeth by Denudation.—3. Swelling of the Fang.—4. Gum Boils.—5. Excrescences from the Gum.—6. Deeply-seated Abscesses in the Jaws.—7. Abscess of the Antrum Maxillare.

CHAP. II.—OF THE DISEASES OF THE ALVEOLAR PROCESSES AND THE CONSEQUENCES OF THEM 199

    III.—OF THE DISEASES OF THE GUMS AND THE CONSEQUENCES OF THEM 204

1. The Scurvy of the Gums, vulgarly so-called.—2. Callous Thickenings of the Gums.

CHAP. IV.—OF NERVOUS PAINS IN THE JAWS . . . 208
    V.—OF THE EXTRANEOUS MATTER UPON THE TEETH . 210
    VI.—OF THE IRREGULARITY OF THE TEETH . . 215
    VII.—OF IRREGULARITIES BETWEEN THE TEETH AND JAW . 221
Of Supernumerary Teeth 221
    VIII.—OF THE UNDER JAW 222
    IX.—OF DRAWING THE TEETH 224

Transplanting Teeth—Of the State of the Gums and Sockets—Age of the Person who is to have the Scion Tooth—the Scion Tooth—Replacing a Sound Tooth when drawn by mistake—Transplanting a Dead Tooth—The Immediate Fastening of a Transplanted Tooth.

CHAP. X.—OF DENTITION . . . 242
Of the Cure—Of Cutting the Gums—Cases.

INDEX.

Abscess of Antrum, 196, N. *w*, 198.
Alveolar Abscess, N. *s*, 192.
    Processes, 3, 4, 7.
        Characters of, in African Negro, N. *i*, 5.
        Diseases of, 199.
        Formation of, 110, N. *s*, 111.
Antrum Maxillare, Abscess of, 196, N. *w*, 198.

Bicuspides, 78.
Bone, 34.
Bony Part of a Tooth, 44.

Canines, Comparative Anatomy of, N. *g*, 70.
    Uses of, N. *i*, 77.
Cases, 249.
Cavity of the Teeth, 53.
Cement, N. *l*, 34.
    Composition of, N. *l*, 37.
Condyle, 6.
    Characteristics of, in Mammalia, N. *j*, 6.
Coronoid Process, 6.
Crusta Petrosa, N. *l*, 34,
Cuspidatus, 70.

Decay of the Teeth, 161.
        By Denudation, 181, N. *o*, 183.
        Theories respecting, N. *e*, 166.
Dental Pulps or Bulbs, Origin of, N. *v*, 115.
Dental Tissues, N. *l*, 34.
    Canals, N. *q*, 45
Dentinal Pulp, N. *v*, 54.
    Uses of, N. *u*, 53.
Dentine, N. *l*, 34.
    Composition of, N. *r*, 50.
    Structure of, N. *q*, 45.
Dentition, 242.
    Cause of Pain in, 120.
Digastricus, 26.
    Comparative Anatomy of, N. *d*, 27.
Diseased Teeth, Changes in, 161, N. *a*, 161.
        Treatment of, 175.

Enamel, 34, 40, N. n, 40.
    Composition of, N. p, 43.
    Formation of, 135, N. j, 126.
Epulis, N. z, 207.
Exostosis, N. s, 51.
Extraction of Teeth, 224.
    Hæmorrhage after, 226, N. d, 227.

Fang, Swelling of, 184.

Glenoid Cavity, 9, N. p, 9.
    Boundaries of, 10.
    Comparative Anatomy of, N. p, 9.
Grinders, 81.
    Comparative Anatomy of, N. o, 81.
Gum Boil, 186.
    Treatment of, 190.
Gums, 106, N. r, 106.
    Diseases of, 204.
    Excrescences from, 193.
    Lancing of, 247, N. g, 247.
    Thickenings of, 206.

Incisores, 61, N. d, 61.
    Comparative Anatomy of, N. d, 61.
Interarticular Cartilage, 11, N. s, 11.
Intermaxillary bones, N. a, 1.

Jaw Bone, Lower, 3.
    Upper 1.
Jaws, Abscess in, 194.
    Growth of, 141.
    Movements of, in Young and Old People, 109.
    Nervous Pains in, 208.

Lower Jaw, 3.
    Articulation of, 9.
    Interarticular Cartilage of, 11.
    Motion in the Joint of, 13.
    Muscles of, 16.
    Ossification of, 4.

Masseter Muscle, 17.
    Comparative Anatomy of, N. y, 18.
Maxillary Bones, Superior, N. a, 1.

Neuralgia, N. a, 209.

Osteo-Dentine, N. l, 34.

Palate, Breadth of, 3.
Part II.—Introduction, 159.
Periosteum of the Teeth, 55.
Premaxillary Bones, N. a, 1.
Premolars, N. j, 78.
    Comparative Anatomy of, N. o, 81.
Pterygoid Muscles, 22.
    Comparative Anatomy of, N. b, 26.

Pterygoideus Externus, 22.
            Internus, 22.

Scurvy in the Gums, 204, N. y, 206.
Smith, Dr. John, on the Temporo-Maxillary Joint, N. r, 14.
Sound Teeth, Replacement of, 234, N. e, 236.
Spina Ventosa, 51, 184, N. p, 185.
Stopping Teeth, 178.
            Material used for, 179, N. l, 180.
Suture-Maxillo-intermaxillary, N. a, 1.

Tartar, N. r, 155.
            Composition of, N. b, 214.
Teeth, Action of, 107.
            Articulation of, 105.
            Characters of, in Man, 5, 6.
            Cleaning of, 154.
            Continual Growth of, 146.
            Decay of, 161.
            Diseases of, 153, 159, 161.
            Drawing of, 224.
            Extraneous Matter upon, 210.
            Formation of, 115.
            Formula of, in Man, N. b, 59.
            Human, in Relation to Voice, 151.
                        Under what Class they come, 152.
            Inflammation of, 171.
            Irregularity of, 215.
            Liability to Decay, N. r, 188.
            Names of, 58.
            Number of, 57, N. a, 57.
            Permanent Formation of, 121, 127.
                        Eruption of, N. c, 125.
                        Sensibility of, 149.
                        Shedding of, 146.
                        Situation of, 56, N. z, 56.
                        Stopping of, 178.
            Supernumerary, 149, 221, N. o, 150.
            Temporary, when cut, N. y, 120.
            Transplanting of, 156, 230, N. s, 157.
Teeth and Jaw, Irregularities between, 221.
Temporal Muscle, 19.
                        Comparative Anatomy of, N. z, 21.
Temporo-Maxillary Articulation, 13, N. w, 13.
Tooth Cavity, Filling up of, 145, N. m, 145.
            Definition of, 34.
            Ossification of, 129, N. i, 129.
            Structure of, 34.
Transplanted Teeth, Fostering of, 237.
Transplanting of Teeth, 156, 230, N. s, 157, f. 241.
            Dead Teeth, 237.

Under Jaw, 222.

Vaso-dentine, 34.

Wisdom Teeth, Cutting of, 248, N. h, 248.

ERRATA.

PAGE 102 —24th line from top—for p. 192, Vol. III., read p. 35.
,, 122 —4th line from bottom—for note v, p., read note v, p. 115.
,, 123—8th line from bottom—for note v, p. 337, read note y, p. 55.
,, 141—last line—for p. 343, read p. 121.
,, 184—at foot note—for p. 37, read p. 51.
,, 197—15th line from top —for finus, read sinus.

PART THE FIRST.

THE NATURAL HISTORY OF THE HUMAN TEETH.

OF THE UPPER JAW.

BEFORE we enter into a description of the Teeth themselves, it will be necessary to give an account of the Upper and Lower Jaw-bones, in which they are inserted; insisting minutely on those parts which are connected with the Teeth, or serve for their motion and action, and passing over the others slightly.

The Upper-Jaw is composed of two bones, (*a*) which generally

---

(*a*) [The superior maxillary bones (*maxillæ superiores*). In addition, it may be stated that the palatine bone on each side is generally reckoned as constituting part of the upper jaw, although in the cavity of the mouth it is confined to the palate, and forms no portion of the alveolar process. In the class Mammalia, the upper jaw consists of three bones on each side,—the palatine, superior maxillary, and inter- or pre-maxillary bones. The last-named bones constitute the anterior portion of the jaw, and form the alveoli of the superior incisors. Whatever modification in size or shape the teeth implanted in the premaxillary bones may display, they are, nevertheless, classed as incisors in the dental formulæ of naturalists. Anchylosis between the superior and premaxillaries takes place sooner or later in life in the typical Quadrumana. In the Chimpanzee, the facial portion of the maxillo-inter-maxillary suture disappears at or about the period of the first dentition; in the Orangs, the suture remains until the development of the great canine teeth; in the Semnopitheci and Inui, the separation continues

remain distinct through life. (*b*) They are very irregular at their posterior and upper parts, sending upwards and backwards a

distinct until the period of old age. In Man, on the contrary, the process of anchylosis (with the exception of a portion of the suture on the palate, which usually remains until after birth) is completed before the termination of intra-uterine existence. It is, therefore, only in the foetal condition that Man can be said to possess an intermaxillary bone; and even this has been denied by many writers, on the ground that the fissure of separation cannot be traced through the external alveolar plate. On the other hand, analogy, the constant existence of the nasal and palatal portions of the maxillo-intermaxillary fissure in intra-uterine life, and the entire separation of the incisive portion of the superior maxillary bone in certain cases of deformity (double hare-lip), are strong arguments for admitting the premaxillary as an original element of the human skull. Direct observation appears, however, to have definitively settled the question. Some years since Dr. Leidy announced to the Academy of Natural Sciences of Philadelphia (1) that he had found the intermaxillary entirely separable in an embryo of one inch eleven lines in length. The division in this case was traceable through the alveolar ridge at a point corresponding to the separation between the incisor and canine alveoli. Figures of the preparation taken from those published in the Proceedings of the Academy will be found amongst the Engravings which illustrate this work. The more recent researches (2) of Robin and Magitot strongly confirm Dr. Leidy's observation. These observers have discovered the existence of a separate formative cartilage for each premaxillary, in which an osseous point is developed two or three days after the first deposition of ossific matter in the alveolar edge of the formative cartilage of the superior maxillary.

It is stated by Soemmerring, that vestiges of the maxillo-intermaxillary suture are to be noticed in Negro crania. Dr. Prichard long since denied any constancy of difference between the melanous and leucous varieties in this respect, and extended observation will verify his statement. In skulls of all nations, it is not very uncommon to find some small vestige of the suture both on the palate and in the floor of the nares, extending outwards from the anterior palatine canal; more rarely it may be traced on the palate to within a short distance of the alveolus. An instance of the permanency of any portion of the suture on the face has never fallen under the Editor's notice.]

(*b*) [Anchylosis between the superior maxillary bones is exceptional in

---

(1) Proceed. Acad. Nat. Sciences, Philadelphia, Jan., 1849, p. 145.

(2) *Vide* Journal de la Physiologie. Edited by Dr. Brown-Séquard. Janvier, 1860, p. 6.

great many processes, that are connected with the bones of the Face and Skull. The lower and anterior parts of the Upper-Jaw are more uniform, making a kind of a circular sweep from side to side, the convexity of which is turned forwards; the lower part terminates in a thick edge, full of sockets for the Teeth. This edge is called in each bone the Alveolar Process. Behind the Alveolar Processes there are two horizontal lamellæ, which uniting together, form part of the roof of the Mouth, which is the partition between the Mouth and the Nose.

This plate, or partition, is situated about half an inch higher than the lower edge of the Alveolar Process; and this gives the roof of the Mouth a considerable hollowness.(c)

The use of the Upper-Jaw is to form part of the Parietes of the Mouth, Nose, and Orbits; to give a basis, or supply the Alveolar Process, for the superior row of Teeth, and to counteract the Lower-Jaw; but it has no motion itself upon the bones of the Head and Face.(d)

OF THE LOWER-JAW.

As the Lower-Jaw is extremely moveable, and its motion is indispensably necessary in all the various operations of the

---

Man. Obliteration of the facial portion of the mesian premaxillary suture commonly takes place as life advances in the Anthropoid Apes.]

(c) [The hollowness and breadth of the palate vary in individual crania. It will be also found of greater comparative length in prognathic skulls. In the larger number of Australian crania, and in those of other dark-skinned races which approach the Australo-Tasmanian type, the bony palate is relatively of great breadth and length.]

(d) [The fixity of the bones constituting the upper jaw is a character common to Mammalia, and contrasts with the moveable condition of the upper mandible in birds—the distensibility resulting from the elastic ligamentous connection which subsists between the premaxillary and maxillary bones in the constricting serpents—the moveable articulation of the superior maxillary in the poisonous serpents—and with the loose connection of the corresponding bones in fishes, which permits, in addition to the ordinary movements, those of protrusion and retraction.]

Teeth, it requires to be more particularly described. It is much more simple in its form than the Upper, having fewer processes, and these not so irregular. Its anterior circular part is placed directly under that of the Upper-Jaw; but its other parts extend farther backwards.

This Jaw is at first composed of two distinct bones; (*e*) but these, soon after birth, unite into one, at the middle of the chin. (*f*) This union is called the Symphysis of the Jaw. Upon the upper edge of the body of the bone is placed the Alveolar Process, a good deal similar to that of the Upper-Jaw. The Alveolar Process extends all round the upper part of the bone,

---

(*e*) [The ossification of the lower jaw, from two ossific centres, each of which forms a lateral half, is one of the few osteological characters which Professor Owen enumerates as being common and peculiar to the Mammalia. (1) In the human subject, authorities are at variance as to the ossification of the inferior maxilla. Kerckringius describes a separate ossific centre for the coronoid process; and he is confirmed by Autenrieth, Spix, and Béclard. Autenrieth enumerates two other separate points of ossification—one in the condyle, the other in the angle. Another, forming the inner border of the alveolus, is described and figured by Spix, who also admits those mentioned by Autenrieth. Cruveilhier allows the separate ossification of the internal alveolar plate, but denies the existence of other secondary nuclei. On the other hand, J. F. Meckel and Nesbitt maintain the ossification of each lateral half from a single centre. The former anatomist attributes the appearance of the separation of the internal alveolar plate to the depth of the groove or fissure for the mylo-hyoid nerve. One reason for the discrepancy which prevails with regard to the ossification of the mandible, is the rapidity with which the process is carried on, and the early period at which it commences. Bony deposit is found in the inferior maxillary earlier than in any other bone except the clavicle. (2)]

(*f*) [Union of the lateral halves takes place during the first year after birth, but a trace of separation may be found at the upper part in the beginning of the second year. (3)]

---

(1) On the Classification and Geographical Distribution of the Mammalia, p. 13.

(2) Quain's Anat., 6th edition, vol. i., p. 78.

(3) Op. cit., p. 78.

from the Coronoide Process of one side, to that of the other.(*g*)
In both Jaws they are every where relatively proportional to the
Teeth; being thicker behind, where the Teeth are larger,(*h*) and
more irregular, upon account of the more numerous fangs inserted into them.(*i*) The Teeth that are situated backwards, in

---

(*g*) [Unbroken continuity of the alveolar series in both jaws is amongst
existing Mammalia only to be found constantly in Man. In the highest
genus of the Quadrumana (*Troglodytes*), there is a considerable *diastema*
or interspace between the canine and lateral incisor in the upper jaw, and
usually a smaller interval is present between the canine and anterior
premolar in the lower. In the larger species of Orang (*Pithecus Satyrus*),
a corresponding wide interval exists between the upper canine and the
contiguous incisor; in the lower jaw the occurrence of a *diastema* is
not constant. In one specimen of the smaller species of Orang
(*Pith. Morio*), described by Professor Owen, the upper jaw exhibited
the ordinary *diastema;* but in a second specimen it was absent, and the
teeth were as uninterrupted as in the human subject. In both specimens,
the implantation of the teeth in the lower jaw was continuous. In
Pithecus Morio, the occasional absence of interval in the alveolar series
would appear to depend on the large size of the teeth, which equal
those of the Great Orang and the small relative proportions of the
maxillæ. In Man, the constant continuity of the alveolar series is
evidently associated with the equable length of the teeth, the shortness
of the jaws, and the small size of the cuspidati. In the Great Apes, when
the mouth is closed, the long crown of the laniary is lodged in the
corresponding interspace between the teeth of the opposed jaw. The
value, however, of this as a peculiar characteristic of Man is diminished
by the discovery of a similar unbroken proximity of the teeth in some
extinct Quadrupeds; *e.g.*, Anoplotherium, Nesodon, Dichodon. (1)]

(*h*) [The alveoli in Man increase regularly in size from the incisors to
the true molars. In the Anthropoid Apes, the largest socket is that for
the great canine tooth.]

(*i*) [In the greater number of skulls of the Australo-Tasmanian variety
of Man, and in some skulls of African Negroes, especially of those in-
habiting the Western Coast, the alveoli of the grinding series will be
found of great relative breadth. This thickness of the molar alveoli in
these races is dependent on the large size and complex implantation of
the corresponding teeth.]

---

(1) Owen, Transactions of the Zool. Soc.; Art. "Teeth," Todd's Cyclopædia
of Anatomy; Quarterly Journal of the Geol. Soc., February 1848.

the Upper-Jaw, have more fangs than those that correspond with them in the lower, and the sockets are accordingly more irregular. The Alveolar Process of the Upper-Jaw, is a section of a larger circle than that of the lower, especially when the Teeth are in the sockets. This arises chiefly from the anterior Teeth in the Upper-Jaw being broader and flatter than those in the Lower. The posterior part of the bone on each side rises almost perpendicularly, and terminates above in two processes; the anterior of which is the highest, is thin and pointed, and is called the Coronoide Process. The anterior edge of this process forms a ridge, which goes obliquely downward and forward on the Jaw, upon the outside of the posterior sockets. To this process the Temporal Muscle is attached; and as it rises above the center of motion, that Muscle acts with nearly equal advantage in all the different situations of the Jaw.

The Posterior Process, which is made for a moveable articulation with the head, runs upward, and a little backward; is narrower, thicker, and shorter, than the anterior; and terminates in an oblong rounded head,(*j*) or Condyle, whose longest axis is nearly transverse. The Condyle is bended a little forward; is rounded, or convex, from the fore to the back part; and likewise a little rounded from one end to the other, or from right to left.(*k*) Its external end is turned a little forward, and

---

(*j*) [In all animals which suckle their young, the condyle is convex or flat, never concave. In birds and reptiles, the articular surface is concave. These characters are of the greatest value to the palæontologist in the determination of fossil remains.]

(*k*) [The shape of the articulating condyle in the human mandible contrasts with the configuration of the same process in the highest known Ape (*Troglodytes Gorilla*). In Man the thickest and most prominent part of the process is near the middle of the joint. In the Gorilla, the inner end of the articulating condyle is the larger, and the posterior border of the articulating surface wants that abrupt definition which is usually observed in Man. In this latter respect the Chimpanzee (*Trog. niger*) makes a nearer approach to the human conformation. (1)]

---

(1) *Vide* Owen on the Osteology of the Chimpanzees and Orangs, Zool. Trans., vol. iv., p. 89.

its internal a little backward; so that the axis of the two Condyles are neither in the same straight line, nor parallel to each other; but the axis of each Condyle, if continued backwards, would meet, and form an angle of about one hundred and forty-six degrees; and lines drawn from the Symphysis of the Chin, to the middle of the Condyle, would intersect their longest axis, at nearly right angles. There are, however, some exceptions; for in a Lower-Jaw, of which I have a drawing, the angle formed by the supposed continuation of the two axes, instead of being an angle of one hundred and forty-six degrees, is of one hundred and ten only. The Lower-Jaw serves for a base to support the Teeth in the Alveolar Process, during their action on those of the Upper-Jaw in mastication; and to give origin to some muscles that belong to other parts.

OF THE ALVEOLAR PROCESSES.

The Alveolar Processes are composed of two thin bony plates, one external, and the other internal. These two plates are at a greater distance from each other at their posterior ends, than at the anterior, or middle part of the Jaw. They are united together by thin bony partitions going across, which divide the processes at the anterior part, into just as many distinct sockets as there are Teeth; but at the posterior part, where the Teeth have more than one root or fang, there are distinct cells, or sockets, for every root. These transverse partitions are more protuberant than the Alveolar Plates; and thus add laterally to the depth of the Cells, particularly at the anterior part of the Jaw. At each partition, the external plate of the Alveolar Process is depressed, and forming furrows, or a fluting rounds the cells, or cavities, for the roots of the Teeth. This is observable in the whole length of the Alveolar Process of the Upper-Jaw; and in the fore-part particularly of the Lower-Jaw. The Alveolar Process of each Jaw, form about one-half of a circular, or rather of an elliptical figure; (*l*) and at the fore-part

---

(*l*) [The elliptical figure described by the alveolar borders of both jaws,

in the Lower-Jaw they are perpendicular, (*m*) but project inwards at the posterior part, and describe a smaller circle than the body of the bone upon which they stand; as we shall observe more particularly hereafter, when we come to treat of the Jaws of Old People.

The Alveolar Processes of both Jaws should rather be considered as belonging to the Teeth, than as parts of the Jaws; for they begin to be formed with the Teeth, keep pace with them in their growth, and decay, and entirely disappear, when the Teeth fall; so that, if we had no Teeth, it is likely we should not only have no sockets, but not even these processes, in which the sockets are formed; and the Jaws can perform their motions, and give origin to muscles, without either the Teeth, or Alveolar Processes. In short, there is such a mutual dependence of the Teeth, and Alveolar Processes on each other, that the destruction of the one seems to be always attended with that of the other. (*n*)

---

and the regularly convex horse-shoe shape of the lower-jaw, strongly characterize the human as opposed to quadrumanous configuration. In Man, the molar series, including the premolars, describes a well-marked curve with the convexity outwards. In the Anthropoid Apes, the teeth from the last molar to the canine are arranged almost in a straight line, with a very slight inflection inwards. They are parallel on the two sides, and are joined at right angles by the line of the incisive alveoli, which unites them in front. (1)

(*m*) [The direction of the incisor alveoli varies in individuals and in races. In the melanous varieties, it is usual to find the anterior portion of the alveolar border projecting in both jaws, especially in the upper; in the leucus and xanthous varieties, the anterior contour of the incisor alveoli is more usually perpendicular. But individuals occur in every variety in whom a greater or less degree of the prognathic configuration is discernible.]

(*n*) [This observation is strictly correct: however rapidly the gum becomes absorbed, whether from indigestion, the use of mercury, the accumulation of calcareous matter, or that affection which is vulgarly termed scurvy in the gum, the alveolar process never becomes exposed (unless it be a dead portion exfoliating), but absorption of the bone always keeps pace with that of the gum.—T. Bell.

---

(1) Op. cit.

In the head of a young subject which I examined, I found that the two first Incisor Teeth in the Upper-Jaw had not cut the Gum; nor had they any root or fang, excepting so much as was necessary to fasten them to the Gum, on their upper surface; and on examining the Jaw, I found there was no Alveolar Process, nor sockets, in that part. What had been the cause of this, I will not pretend to say: whether it was owing to the Teeth forming not in the Jaw, but in the Gum; or to the wasting of the fangs. The appearance of the Tooth favoured the first supposition; for it was not like those, whose fangs are decayed in young subjects, in order to the shedding of the Teeth; and as it did not cut the Gum, it is reasonable to think it never had any fang. That end from which the fang should have grown, was formed into two round and smooth points, having each a small hole leading into the body of the Tooth, which was pretty well formed.

OF THE ARTICULATION OF THE LOWER-JAW.

Just under the beginning of the Zygomatic Process of each Temporal Bone, before the external Meatus Auditorius, an oblong cavity(*o*) may be observed; in direction, length, and breadth, in some measure corresponding with the Condyle of the Lower-Jaw.(*p*) Before, and adjoining to this cavity, there is an oblong

---

(*o*) [In all Mammals the articular surface for the condyle is either concave or flat.]

(*p*) [It is evident that the description in the text refers only to the articulating portion of the glenoid cavity, or that anterior to the Glaserian fissure. The posterior portion of the cavity lodges a process of the parotid gland, and does not enter into the articulation. In Man, as in all other Mammalia, the lower jaw articulates with the squamous element of the temporal bone; in them the tympanic element is solely subservient to the organ of hearing. But this is not the case in the other vertebrate classes. In birds, reptiles, and osseous fishes, the mandible articulates with bones, which are the homologues of the tympanic ring in Man. In cartilaginous fishes, the articulating surface is formed by the pterygoid bone, the homologue of the internal pterygoid plate in the human subject. In the Lepidosiren, which in other re-

eminence,(*q*) placed in the same direction, convex upon the top, in the direction of its shorter axis, which runs from behind forwards; and a little concave in the direction of its longer axis, which runs from within outwards. It is a little broader at its outer extremity; as the outer corresponding end of the Condyle describes a larger circle in its motion than the inner. The surface of the cavity, and eminence, is covered with one continued smooth cartilaginous crust,(*r*) which is somewhat ligamentous, for by putrefaction it peels off, like a membrane, with the common Periosteum. Both the cavity and eminence serve for the motion of the Condyle of the Lower-Jaw. The surface of the cavity is directed downward; that of the eminence downward and backward, in such a manner that a transverse section of both would represent the Italick letter *S*. Though the emi-

---

spects presents a curious association of the characters of the osseous and cartilaginous fishes, the pterygoid bone contributes the inner, the tympanic the outer portion of the recipient articular surface. (1)

The articulating portion of the glenoid cavity is bounded posteriorly in the Anthropoid Apes, and generally in the lower Mammalia, by a prominent ridge or process. In the Rodentia, a similar ridge bounds the articulating cavity internally. In Man this posterior boundary is absent, or but slightly indicated. The downward development of the human cranium posterior to the articulating surface, affording, as it does, a support against backward dislocation, obviates the necessity for the development of a post-glenoid process. (2) When its rudiment exists in Man, it is known as the middle root of the zygoma. I believe that the Author of the article Temporo-Maxillary Articulation is mistaken when he asserts that this process is more frequently indicated in the lower races of mankind. I have seen it at least as frequently in the skulls of the natives of Europe and Asia as in African and Australian crania.]

(*q*) [Anterior root of the zygoma.]

(*r*) [It is only that portion of the glenoid cavity concerned in the articulation of the joint which is covered with cartilage.]

---

(1) Owen, Trans. Lin. Soc., vol. xviii., p. 336. Article Temporo-Maxillary Articulation, in Todd's Cyclopædia of Anatomy, by S. R. Pittard.

(2) Owen, Zool. Trans., vol. i., p. 346.

hence may, on a first view of it, appear to project considerably below the cavity, yet a line drawn from the bottom of the cavity, to the most depending part of the eminence, is almost horizontal, and therefore nearly parallel with the line made by the grinding surfaces of the Teeth in the Upper-Jaw: and when we consider the Articulation farther, we shall find that these two lines are so nearly parallel, that the Condyle moves almost directly forwards, in passing from the cavity to the eminence; and the parallelism of the motion is also preserved by the shape of an intermediate cartilage.

In this joint there is a moveable cartilage,(*s*) which, though common to both Condyle and cavity, ought to be considered rather as an appendage of the former than of the latter, being more closely connected with it; so as to accompany it in its motion along the common surface of both the cavity and eminence. This cartilage is nearly of the same dimensions with the Condyle, which it covers; is hollowed on its inferior surface, to receive the Condyle: (*t*) on its upper surface it is more unequal, being moulded to the cavity and eminence of the articulating surface of the Temporal Bone, though it is considerably less, and is therefore capable of being moved with the Condyle, from one part of that surface to another. Its texture is ligamento-cartilagineous. This moveable cartilage is connected with both the Condyle of the Jaw, and the articulating surface of the Temporal Bone, by distinct ligaments, arising from its edges all round.(*u*) That by which it is attached to the

---

(*s*) [The inter-articular fibro-cartilage. It is the only inter-articular fibro-cartilage into which muscular fibres are inserted—viz., a portion of the external pterygoid, which it receives along its anterior border, and which is instrumental in moving the fibro-cartilage in the antero-posterior motion of the condyle. The inter-articular fibro-cartilage is constantly found in animals which suckle their young, but not in the inferior vertebrata.]

(*t*) [The inter-articular fibro-cartilage is thicker at the edges than in the centre; it is not unfrequently perforated in the latter situation.]

(*u*) [Thin and short ligamentous fibres (*membrana articularis*,

Temporal Bone, is the most free and loose; though both ligaments will allow an easy motion, or sliding of the cartilage on the respective surfaces of the Condyle, and Temporal Bone. These attachments of the cartilage are strengthened, and the whole articulation secured, by an external ligament, which is common to both, and which is fixed to the Temporal Bone, and to the neck of the Condyle. On the inner surface of the ligament, which attaches the cartilage to the Temporal Bone, and backwards, in the cavity, is placed what is commonly called the Gland of the Joint; at least, the ligament is there much more vascular than at any other part. (*v*)

Weitbrecht) surround the greater portion of the joint, cover the synovial membranes, and serve to connect the fibro-cartilage with the osseous margins. These thin ligamentous structures are not distinguished in the text from the synovial bursæ, neither is it usual to find the latter noticed specially in anatomical works of the period. They had been, however, described at some length by Weitbrecht, whose work appeared in 1742. (1) They are two in number—one placed above the fibro-cartilage, the other below it. The superior is the larger and looser; it lines the upper surface of the inter-articular cartilage and the smooth part of the glenoid cavity. The inferior is interposed between the condyle and the lower surface of the fibro-cartilage. When the latter structure is perforated in the centre, the synovial bursæ communicate and form one cavity.]

(*v*) [Dr Clopton Havers,(2) in 1691, described the vascular processes of the synovial membrane, which are commonly found projecting more or less into the cavities of joints. When these processes are of any size, they generally contain fat. They are frequently cleft, so as to present a fringed appearance at their free border, which is very vascular. Dr Havers regarded them as special structures for the secretion of synovia, and named them the " mucilaginous glands of the joints." Subsequent anatomists have generally denied them a special function, although, as extensions of the synovial membrane, they must necessarily increase the amount of secretion. Havers's view was revived by Mr Rainey, in a paper published in the Proceedings of the Royal Society, in 1846. He bases his advocacy on the constant occurrence of such vascular processes, not only in the joints, but also in synovial sheaths, on a peculiar convoluted condition of their blood-vessels, and on the arrangement of the epithelium covering them, which, " besides enclosing separately each

---

(1) Syndesmologia, p. 80, 1742.
(2) Osteologia Nova, p. 187, 1691.

OF THE MOTION IN THE JOINT OF THE LOWER-JAW.

The Lower-Jaw, from the manner of its articulation, is susceptible of a great many motions. (*w*) The whole Jaw may be brought horizontally forwards, by the Condyles sliding from the cavity towards the eminences on each side. This motion is performed chiefly when the Teeth of the Lower-Jaw are brought directly under those of the Upper, in order to bite, or hold any thing very fast between them.

packet of convoluted vessels, sends off from each tubular sheath secondary processes of various shapes, into which no blood-vessels enter." Mr Rainey's account of the structure of these processes has since been confirmed in most particulars by Kölliker.(1)]

(*w*) [The form of the osseous surfaces entering into the temporo-maxillary articulation, and the diversities of motion which it admits, are so modified to suit the functions and necessities of the different families of Mammalia, that there is no portion of the skeleton calculated to yield more certain information to the naturalist. In the frugivorous Quadrumana the articulation is loose, as in Man, and permits within certain limits each kind of movement—viz., in the vertical, the antero-posterior, and lateral directions. The principal differences between bimanous and quadrumanous structure in this joint are the somewhat greater flatness of the glenoid surface, the minor development of the anterior articular eminence, and the constancy and size of the post-glenoid ridge in the latter. In the Carnivora the condyle of the lower jaw is an oblong cylindroid process of considerable length, placed transversely, almost, if not quite, in a straight line with its fellow of the opposite side. The receiving surface is bounded anteriorly and posteriorly by salient ridges, which increase its depth, and limit the chief motion of the jaw to the vertical direction, permitting only in addition a slight lateral gliding of the condyles. The development of the anterior ridge varies in different species. In the badger both processes project in such a manner as almost to embrace the condyle, so that in the skeleton the inferior maxilla is retained *in situ* without any artificial fastening. In the placental Rodents (the Hares excepted), on the other hand, the glenoid cavity is a deep groove excavated longitudinally under the base of the zygoma, and corresponds with the long diameter of the usually oval condyle, which is set antero-posteriorly instead of from side to side. In accordance with this conformation, the conspicuous movement of the lower jaw in the Rodents is in the antero-posterior direction. The "nibbling" motion allowed is in

---

(1) *Vide* Quain's Anat., sixth edit., vol. i., p. ccxlix.

Or, the Condyles only may be brought forwards, while the rest of the Jaw is tilted backwards, as in the case when the Mouth is open; for on that occasion the angle of the Jaw is tilted backwards, and the chin moves downwards, and a little backwards also. In this last motion, the Condyle turns its face a little forwards; and the center of motion lies a little below the Condyle, in the line between it and the angle of the Jaw. By such an advancement of the Condyles forwards, together with the rotation mentioned, the aperture of the Mouth may be considerably enlarged; a circumstance necessary on many obvious occasions.

The Condyles may also slide alternately backwards and forwards, from the cavity to the eminence, and *vice versa;* so that while one Condyle advances, the other moves backwards, turning the body of the Jaw from side to side, and thus grinding, between the Teeth, the morsel separated from the larger mass by the motion first described. (*x*) In this case, the center of motion lies exactly in the middle between the two Condyles. And it is to be observed, that in these slidings of the Condyles forwards

---

exact conformity with the structure of their teeth and the disposition of the muscles of mastication. In the Ruminantia, again, we find both glenoid surface and condyle almost flat: the former is bounded behind by a transverse crest which passes inwards from the zygoma, but has no corresponding anterior limitation. Such a joint is admirably adapted for an extensive lateral movement of the jaw in chewing the cud, whilst the power of gaping is proportionately limited. (1)]

(*x*) [In a paper read before the Royal Society of Edinburgh, Dr John Smith has advanced the following theory of the mechanism of the Temporo-Maxillary Joint. He considers that in the movement of simply opening and shutting the mouth, "the condyles cannot act as a *simple hinge*, as they lie, not at right angles to the plane of motion of the lower jaw, but obliquely to it, each condyle looking inwards and forwards. Their more perfect action, therefore, cannot occur in this movement, but seems to belong to that of mastication. The articulating surface, strictly speaking, on each condyle appears to constitute the

---

(1) *Vide* Cuvier, Leçons d'Anatomie Comparée, Tome iv., P. 1, p. 36, edit. 1835; Todd's Cyclopædia of Anatomy, Article Temporo-Maxillary Articulation.

and backwards, the moveable cartilages do not accompany the Condyles in the whole extent of their motion; but only so far as to adapt their surfaces to the different inequalities of the Temporal Bone: for as these cartilages are hollow on their lower surfaces where they receive the Condyle, and on their opposite upper surfaces are convex where they lie in the cavity; but forwards, at the root of the eminence, that upper surface is a little hollowed; if they accompanied the Condyles through the whole extent of their motion, the eminences would be applied to the eminences, the cavities would not be filled up, and the whole articulation would be rendered very insecure.

This account of the motion of the Lower-Jaw, and its cartilages, clearly demonstrates the principal use of these cartilages; namely, the security of the articulation; the surfaces of the cartilage accommodating themselves to the different inequalities, in the various and free motions of this joint. This cartilage is also very serviceable for preventing the parts from being hurt by the friction; a circumstance necessary to be guarded against,

---

thread, or rather part of the thread, of a conical screw passing over an axis lying at or about right angles to the plane of motion in simple opening and closing of the jaws. This spiral course of the articular surface is perhaps best seen in some of the large *Carnivora*, such as the lion, but is also obvious in a well-developed human condyle.

"The action of this conical screw or tap within the glenoid cavity, considered as the conical die, takes place with accuracy only when one joint alone acts with the condyle within the glenoid cavity—the other condyle being beyond it, and gliding on the surface of the zygoma, as during mastication. The food is in this process crushed between the molar teeth of that side whose condyle remains within the glenoid cavity, this condyle screwing the jaw back, so to speak, to its natural position at each closure of the teeth.

"By this construction a great amount of friction is avoided; what would otherwise be a *rubbing* being thus converted into a *rolling* motion between the condyloid and glenoid surfaces; while, by one or other condyle always remaining in the glenoid cavity during mastication, greater steadiness and security is afforded to the joint." (1)

It need scarcely be remarked here, that however the above explanation

---

(1) Edinb. Philosoph. Journal, N.S., vol. viii., p. 150.

where there is so much motion. Accordingly, I find this cartilage in the different tribes of Carnivorous Animals, where there is no eminence and cavity, nor other apparatus for grinding; and where the motion is of the true ginglimus kind only.

In the Lower-Jaw, as in all the joints of the body, when the motion is carried to its greatest extent, in any direction, the muscles and ligaments are strained, and the person made uneasy. The state, therefore, into which every joint most naturally falls, especially when we are asleep, is nearly in the middle, between the extremes of motion; by which means all the muscles and ligaments are equally relaxed. Thence it is, that commonly, and naturally, the Teeth of the two Jaws are not in contact; nor are the Condyles of the Lower-Jaw so far back in the Temporal Cavities as they can go.

OF THE MUSCLES OF THE LOWER-JAW.

Having described the figure, Articulation, Motion, and use of the Lower-Jaw, it will be necessary, in the next place, to give some account of the Muscles that are the causes of its motion.

There are five pair of Muscles, each of them capable of producing various motions, according to the situation of the Lower-Jaw, whether they act singly, or in conjunction with others; and two or more of them may be so situated, as to be capable of moving the Jaw in the same direction; and every motion is produced by the action of more than one Muscle at a time. Thus, if the Jaw is depressed, and brought to one side, either the Masseter, Temporal, or Pterygoidæus internus of the opposite

---

may be applicable to the movements of the condyle in the human subject, it cannot be extended to the motion of the joint in the large Carnivora; in which animals, nevertheless, a spiral direction of the articular surface is well seen. This spiral appearance of the articular surface in the *Felidæ* is clearly dependant on the disposition of the anterior and posterior articular processes, the post-glenoid being most developed internally, whilst the anterior ridge is necessarily more external to permit the play of the contiguous coronoid process.]

side will not only raise the Jaw, but bring it to its middle state. It will be necessary in the description of each Muscle, to give its use in the different situations of the Jaw; by which means, after they are all described, their compound actions will be better understood. I shall first describe those which raise the Jaw; then those which give it the lateral motion; and lastly, those which depress it; proceeding in each class as they rise in dissection.

The most superficial is the Masseter: it is situated upon the posterior and lower part of the Face, between the cheek-bone, and angle of the Lower-Jaw, directly before the lower part of the Ear. It is a thick, short, complex Muscle, and a little flattened: it appears to have two distinct origins, an anterior outer, and a posterior inner; but that is owing only to its outer edge at its origin being slit, or double; and the fibres of these two edges having a different course, decussating each other a little. The anterior, and outer portion of the Muscle begins to rise from a small part of the lower edge of the Malar Process of the Maxillary Bone, adjoining to the Os Malæ; and continues its origin all along the lower horizontal edge of this last bone, to the angle where its Zygomatic Process turns up, to join that of the Temporal Bone. The external layer of fibres in this portion are tendinous at their beginnings, while the internal are fleshy.

The posterior and inner portion of this Muscle begins to rise partly tendinous, and partly fleshy, from the same lower edge of the Os Malæ; not where the origin of the other portion terminates, but a little farther forwards; and this origin is continued along the lower edge of the Zygomatic Process of the Temporal Bone, as far backwards as the eminence belonging to the articulation of the Lower-Jaw.

From this extent of its origin, the Muscle passes downwards to its insertion into the Lower-Jaw. The anterior external portion is broader at its insertion than at its origin; for it occupies a triangular space of the Lower-Jaw above the angle, and on the outside, of about an inch in size, to about an inch and a half from the angle towards the Chin. In consequence of this extent of insertion, the fibres of this portion divaricate very considerably. They are mostly fleshy at their insertion, a few only being tendinous, particularly those that are inserted backwards. The posterior and inner portion of the Masseter is narrower at its insertion than at its origin; its posterior fibres running forwards, as well as downwards, while its anterior run almost directly downwards. It occupies in its insertion the remaining part of the scabrous surface, above the angle of the Lower-Jaw, which lies between the anterior portion and the two upper processes, viz., the Condyle and Coronoide. As the anterior fibres of this portion rise on the inside of the posterior fibres of the other portion; and as its posterior fibres run forwards as well as downwards, and its anterior run almost directly downwards, while the fibres of the other portion radiate both forwards and backwards; these two portions in some measure decussate, or cross one another. The anterior fibres, which run farthest and lowest down, are tendinous at their insertion, while the posterior and shortest are fleshy.

The use of the whole Muscle is to raise the Lower-Jaw; and when it is brought forwards, the posterior and inner portion will assist in bringing it a little back; so that this Muscle becomes a rotator, if the Jaw happens to be turned to the opposite side.

We may observe, that this Muscle is intermixed with a number of tendinous portions, both at its origin and its insertion; which give rise to a greater number of fleshy fibres, and thereby add to the strength of the Muscle. (*y*)

(*y*) [The masseter muscle exists in all the inferior Mammalia. It may be said generally to vary in size and strength with the amount of resistance it is destined to overcome in the act of mastication. In those animals, also, in which the other elevators of the jaw, especially the temporal, are of small size, the masseter takes on increased development and power (*e. g.*, Rodentia, Ruminantia). The size, strength, and shape of the zygomatic bar, together with the condition of the angle and posteroexternal surface of the inferior maxilla, may be taken as indices of the development and powers of this muscle. In the typical Carnivora, the zygomatic arch is of very great strength and depth; below it is formed

TEMPORALIS.

It is situated on the side of the Head, above, and somewhat before the Ear. It is a pretty broad, flat, and radiated Muscle; broad and thin at its origin; narrow and thick at its insertion; and is covered with a pretty strong Fascia, above the Jugum.

This Fascia is fixed to the bones round the whole circumference of the origin of the Muscle. Above, it is fixed to a smooth white line, that is observable upon the Skull, extending from a little ridge on the lateral part of the Os Frontis, continued across the Parietal Bone, and making a turn towards the Mammillary Process. It is fixed below, to the ridge where the Zygomatic Process begins, just above the Meatus Auditorius; then to the upper edge of the Zygomatic Process itself, and anteriorly to the Os Malæ. This adhesion, anteriorly, above, and posteriorly, gives, as it were, the circumference of the origin of the Temporal Muscle.

---

almost entirely by the malar bone, above by the temporal, so that the suture between them extends obliquely nearly the whole length of the arch. It is arched upwards as well as outwards, the upper margin being convex, the lower concave, and the latter is deeply excavated for the attachment of the muscle. It is to be observed that the more purely carnivorous the animal, the greater is the convexity of the vertical arch described by the zygoma. The short and very thick masseter arising from the vaulted arch above passes downwards, backwards, and inwards, to be inserted into the posterior part of the lower border of the jaw, and into the lower portion and inferior boundary of the deep fossa which occupies its external surface behind the great molar tooth (*carnassière*), whilst the most external of the fibres turn round the inferior margin of the bone, to be fixed in a tendinous *raphé* common to it and the external pterygoid. (1) The postero-internal portion of the muscle which rises in front of the glenoid cavity has a direction downwards and forwards, and is inserted into the external excavated surface of the coronoid. In the Carnivora, the masseter is at its maximum of development and power as an elevator of the lower jaw. With them we may compare another order —Rodentia—in which this muscle has another purpose to fulfil besides that of raising the jaw—viz., the production of that motion from behind forwards for which we have already seen the articular surfaces of the

(1) *Vide* Straus-Durckheim, Anatomie du Chat, T. ii., p. 217.

TEMPORALIS.

This Muscle arises from all the bones of the side of the Head, that are within the line, for insertion of the tendinous Fascia, viz., from the lower and lateral part of the Parietal Bone, from all the squammous portion of the Temporal Bone, from the lower and lateral part of the Os Frontis, from all the Temporal Process of the Os Sphenoides, and often from a process at the lower part of this surface, (which portion, however, is often common to this Muscle, and the Pterygoidæus externus)

condyle and glenoid cavity are specially adapted. In the Rodents, the lower border of the zygoma is convex, the curve of the vertical arch being always in the opposite direction to that which is constant in the Carnivora ; in some species—*e. g.*, the capybara, the agoutis, and pacas—the convexity of the arch descends to a level below that of the superior grinders. The superior maxillary bone contributes also to its formation, the malar being frequently, as it were, suspended between its apophysis and the zygomatic process of the temporal bone. In these animals the masseter is of great strength when compared with the temporal, and it is divided into several distinct portions, which in some species may rank as separate muscles. A large portion of the muscle will be found to arise from the superior maxillary and fore part of the arch, and to pass obliquely backwards, its action being to bring the jaw directly forwards. In many instances a muscular fasciculus commences by a strong tendon under the sub-orbital foramen, and passes nearly directly backwards ; in the agouti this portion of the masseter covers the posterior half of the jaw, and terminates on the internal surface of its posterior border. Other fibres have a more perpendicular arrangement, and in some species—*e. g.*, the rabbits—the fibres of a small portion of the muscle will be found to run in the opposite direction to the larger mass. They assist in moving the jaw upwards and backwards. (1)

In Man and the Quadrumana, the zygomatic bar is nearly transverse. In the latter it is of greater length, in conformity with the relative dimensions of the cranium and face ; it is also somewhat more curved, the upper border being slightly convex behind and concave in front. In the larger Anthropoid Apes (*Tr. Gorilla and Pith. Satyrus*) the massive dimensions of the zygoma, dependant on the great size of the lower jaw and the development of the canine teeth, present a striking contrast with the human configuration. (2)]

---

(1) *Vide* Cuvier, Leçons d'Anatomie Comparée, T. iv., partie 1, leçon xvi., edit. 1835.

(2) *Vide* Owen, Zool. Trans.

and from the posterior surface of the Os Malae. Outwardly, it rises from the inner surface of the Jugum, and from the whole inner surface of the Fascia above described. At this origin, from the Jugum it is not to be distinguished from the Masseter, being there, in fact, one and the same Muscle; and indeed the Masseter is no more than a continuation of the same origin, under the edge of the Jugum; and might properly enough be reckoned the same, both as to its origin and insertion, and in some measure in its use also.

The origin is principally fleshy; and the Muscle passes from it, in general, downwards, and a little forwards, converging, and forming a thin middle tendon. After which the Muscle runs downwards, on the inside of the Jugum, and is inserted into the Coronoid Process of the Lower-Jaw, on both sides tendinous and fleshy, but principally tendinous. It reaches farther down upon the inside of the Coronoid Process, than upon the outer side, where the insertion is continued as low as the body of the bone.

The posterior and inferior edge of this Muscle passes over the root of the Zygomatic Process of the Temporal Bone, as over a pulley, which confines the action of the Muscle to that of raising the Lower-Jaw, more than if its fibres had passed in a direct course from their origin to their insertion.

The use of the Temporal Muscle, in general, is to raise the Lower-Jaw; and as it passes a little forwards to its insertion, it must bring the Condyle at the same time backwards, and so counteract the Pterygoidæus externus of the opposite side; and if both Muscles act, they counteract both the Pterygoidæi, by bringing back the whole of the Jaw. (z)

(z) [The size and development of the temporal muscle (crotaphite) is indexed by the extent and depth of the temporal fossa, the outward or horizontal curve described by the zygoma, together with the height and the excavations on the internal and external surfaces of the coronoid process. As the powerful elevator of the lower jaw, it is *par excellence* the biting muscle; and it may be generally stated that its dimensions and force will be found closely related to the development of the piercing and cutting teeth, especially the canines. In those of the inferior Mammalia

PTERYGOIDÆUS INTERNUS.

It is situated upon the inside of the Lower-Jaw, opposite to the Masseter, which is upon the outside. It is a strong short Muscle, a little flattened, especially at its insertion. It arises tendinous and fleshy from the whole internal surface of the external Ala of the Sphenoid Bone; from the external surface of the internal Ala, near its bottom; from that process of the Os Palati that makes part of the Fossa Pterygoidæa; likewise from the anterior rounded surface of that process, where it is connected to the Os Maxillare superius. From thence the Muscle passes downwards, a little outwards and backwards, and is inserted tendinous and fleshy into the inside of the Lower-Jaw, from the angle, up almost to the groove for the admission of the Maxillary Nerve, where the surface of the bone is remarkably scabrous.

The Use of this Muscle is to raise the Lower Jaw; and from its direction, one would suspect that it would bring the Condyle a little forwards; but this motion is contrary to that of the Lower-Jaw, for it is naturally brought back when raised.

PTERYGOIDÆUS EXTERNUS,

Is situated immediately between the external surface of the external Ala of the Pterygoid Process, and the Condyle of the

---

in which this muscle attains considerable size, it is common to find the depth of the fossa from which it arises increased by the production of its boundaries in the form of osseous ridges or *cristæ*, which spring from the surfaces of the frontal, the parietal, and the occipital bones. The direct relation which exists between the development of the temporal and the condition of the dental apparatus is well illustrated by a comparison of the male and female and immature and adult crania of the large Anthropoid Quadrumana. Amongst these the maximum of extension and depth of the temporal fossa is to be seen in the skull of the adult male Gorilla, where a lofty sagittal crest is produced by the union of the frontal ridges at the coronal suture, and behind is continued into two thick salient occipital cristæ. The same thing is repeated, although in a somewhat inferior degree, in some skulls of the

Lower-Jaw; lying, as it were, horizontally along the basis of the Skull. It is somewhat radiated in some bodies; broad at the origin, and small at the insertion; but the greater part of

---

adult male Orang (*Pith. Satyrus*). In other specimens of this species, the parietal ridges, although considerably developed, are not blended together in the form of a median crest, but are separated by an interval of varying width. These differences in the crania of the male Orangs do not seem to indicate a difference in race; they are rather the indices of age and muscular energy. (1) In the males of both *Tr. Gorilla* and *Pith. Satyrus* the laniaries attain a development almost rivalling their proportions in the more powerful Carnivora. In the females of these species, on the other hand, which are distinguished by the comparatively small size of the canines, the osseous cristæ are proportionately of minor development; the parietal ridges do not unite to form a high sagittal crest, but are continued separately backwards over the upper surface of the cranium, albeit in the female Gorilla they are only divided by a narrow groove. (2) Again, in the smaller species of Chimpanzee (*Tr. niger*) the muscular processes of the male skull are less developed than in *Tr. Gorilla*, the laniaries also attaining smaller dimensions. But in all these great Apes it is only on the completion of the second dentition that the surface for the attachment of the biting muscles becomes thus extended. In the immature Chimpanzees and Orangs, whilst the cranial portion of the skull is comparatively large, the facial part and especially the jaws are but little developed, the surface of the cranium is uninterrupted, and the temporal fossæ lack the depth which they afterwards acquire, with the full attainment of the powers of combat and mastication. It is at this stage of his existence that the young Ape is so eminently anthropoid: but, "as growth proceeds, the milk teeth are shed, the jaws expand, the great canines succeed their diminutive representatives, the biting muscles gain a proportional increase of carneous fibres, their bony fulcra respond to the call for increased surface of attachment, and the sagittal and occipital crests begin to rise." (3)

It is scarcely necessary to say that no such extension of the temporal fossa is ever found in Man; although in the low, uncivilized races, where the teeth and jaws are put to rough work, it is not unusual to find the arched border of the fossa more strongly defined than in the higher

---

(1) *Vide* Owen, Zool. Trans., vol. iv., p. 165.

(2) Descriptive Catalogue of Osteological Series in Museum of R. Col. of Surgeons, vol. ii., p. 803.

(3) Owen, Class. and Distrib. Mammalia, Appendix on Gorilla, p. 76.

it forms a round strong fleshy belly; so that the part that makes it of the radiated kind is thin.

---

varieties. In both Man and the Quadrumana the zygoma is slightly curved outwards; amongst the latter the outward curve is well marked in the Baboons (*Cynocephali*).

It is in the order Carnivora that the temporal acquires its highest development. This is evidenced by the great depth of the fossa, the outward curve of the zygoma, the development of frontal, parietal, and occipital crests, and the height of the coronoid process of the lower jaw, together with its deeply-excavated external surface. The skulls of the hyæna, wolf, and some of the larger kinds of dog—*e. g.*, the mastiff, blood-hound, &c.—are remarkable for the development and height of the sagittal crest, produced by the union of the frontal and parietal ridges. Amongst the *Felidæ*, the single crest is best expressed in the cranium of the lion; in the smaller species the parietal ridges only meet in the posterior part of their extent. On removing the skin from the upper surface of the head in these animals, we find nothing on each side but the great mass of the temporal, which completely covers the superior and lateral regions. In the Cat, the muscle is described by Straus-Durckheim as consisting of three portions—an antero-external of large size, which arises from the anterior part of the strong temporal aponeurosis and the anterior part of the fossa, and two posterior portions, of which the internal is the larger, and occupies the two posterior thirds of the same fossa. The three heads unite to be inserted into the coronoid process of the jaw, their fibres covering its surfaces and borders, being attached on the inner side as low down as the dental foramen, on the outer to the upper portion of its surface, and to an aponeurosis common to this muscle and the masseter. (1) If we contrast with the Carnivora the Ruminants and Rodents, we find a wide difference in the development of the temporal, as indicated by the size and depth of its fossa. In the latter orders this space is narrow and shallow, the muscular ridges are but slightly expressed; and in the Ruminants the zygomatic arch is short, and projects but little. (2) As another instance of the close relation which exists between the development of the temporal muscle and the dental type, we may adduce the Camel. In the Camel, canines are present in both jaws, together with a pair of laniariform incisors and premolars in the upper jaw. Concomitant on this higher type of dentition, we find the skull furnished with a sharp and deep but thin occipital crest, a low but sharp parietal crest,

---

(1) *Vide* Straus-Durckheim, Anatomie Descriptive et Comparative du Chat, T. ii., p. 216.

(2) Cuvier, op. cit., leçon xvi.

The thick and ordinary portion of it arises tendinous and fleshy, from almost the whole external surface of the external Ala of the Pterygoid Process of the Sphenoid Bone, excepting a little bit of the root at the posterior edge; and towards the lower part, it arises a little from the inner surface of that Ala. The thin portion arises from a ridge of the Sphenoid, that is continued from the process towards the Temple, just behind the Foramen Lacerum inferius, which terminates in a little protuberance. This origin is sometimes wanting; and in that case, the Temporal Muscle arises from that protuberance; and very often this origin is common to both. These two origins of this Muscle are sometimes so much separated, as to make it a Biceps.

From these origins the Muscle passes outwards, and a little backwards, converging; that is, the superior fibres passing outwards and backwards, and a little downwards; while the inferior, or larger portion of it, passes a little upward.

It is inserted tendinous and fleshy into a depression on the anterior part of the Condyle and Neck of the Lower-Jaw, upon the inside of that ridge, which is continued from the Coronoide Process. A little portion is likewise inserted into the anterior part of the moveable cartilage of the joint.

When this Muscle acts singly it is a rotator; for it brings the Condyle of the Jaw forwards, and likewise the moveable cartilage, which throws the Chin to the opposite side; but if it acts in conjunction with its fellow of the opposite side, instead of being turned to one side, the whole Jaw is brought forwards, and thus these counteract the Temporal, &c.

These two Muscles generally act alternately; and when they do so, one acts at the time of depression, the other at the time

---

"whilst the zygomatic arches are longer and span across a wider temporal fossa than in the ordinary Ruminants," (1) in which the upper incisors are absent, being replaced by a callous pad, the upper canine is in many instances wanting, whilst the lower canine resembles and forms one of the same series with the lower incisors.]

---

(1) Osteolog. Catalogue R. Col. Surgeons, vol. ii., p. 572.

of elevation; so that these Muscles act, both when the Lower-Jaw is raised, and when it is depressed: yet they do not assist either in raising or depressing it. (*b*)

DIGASTRICUS.

It is situated immediately under, and a little upon the inside of the Lower-Jaw, and outside of the Fauces, extending from the Mastoid Process to the Chin, nearly along the angle made by the Neck and Chin, or Face. The name of this Muscle

---

(*b*) [The pterygoids vary in size and function in the different families of Mammalia. They are usually largely developed in herbivorous and frugivorous animals, where they are in direct relation to the lateral and antero-posterior motions of the jaw in mastication. In the Chimpanzees and Orangs, the force and size of the internal pterygoid is evidenced by the roughness and elevations which exist on the internal surface of the posterior border of the ascending ramus and angle of the jaw. (1) In the Rodents, the internal pterygoid, like the masseter, is employed in drawing forward as well as in raising the mandible. To increase the obliquity of its fibres and the backward extent of surface for their insertion, the angle of the jaw is prolonged posteriorly. The Ruminants, and grass-feeding animals generally, have these muscles of large size. In the typical Carnivora, on the other hand, the principal pterygoid is simply an elevator of the jaw, not a rotator or protrusor. In the Cat, the larger muscle (*external Pterygoid*, Straus-Durckheim; *internal Pterygoid*, Cuvier) is flat and of considerable size; it has its origin from the whole length of the lateral border of the guttural part of the palatine bone, and from its corresponding orbital surface, and passes outwards, backwards, and downwards, to be inserted into the inner surface of the lower border of the jaw, into the tendinous *raphé* common to it and the masseter, and into the inner surface of the coronoid. The smaller slip (*internal Pterygoid*, Straus-Durckheim) is almost blended with the preceding; it is placed behind, and is in part internal to and under the principal muscle. None of the fibres of the pterygoid in the Cat are attached to the condyle or to the interarticular cartilage. These muscles, as do the temporal and masseter, present an admirable instance of adaptation to the type of dentition, and to the structure of the joint. (2)

---

(1) *Vide* Owen, Zool. Trans., vol. iv.
(2) *Vide* Straus-Durckheim, op. et loc. cit.; Cuvier, op. cit., p. 88.

expresses its general shape, as it has two fleshy bellies, and of course a middle tendon. (*d*) Yet some of its anterior belly does not arise from the tendon of the posterior, but from the Fascia, which binds it to the Os Hyoides. These two fleshy bellies do not run in the same line, but form an angle, just where the tendon runs into the anterior belly; so that this tendon seems rather to belong to the posterior, which is the thickest and longest.

This Muscle arises from the Sulcus made by the inside of the Mastoid Process, and a ridge upon the Temporal Bone, where it

---

(*d*) [The name "digastricus" is not equally applicable to this muscle in all the Mammalia in which it is found. In the Apes, as in Man, it consists of two bellies, united by a middle tendon, which passes through the stylo-hyoid. But the digastric of the higher Quadrumana differs from the same muscle in Man in the greater size and strength of the anterior belly. Duvernoy describes this portion of the muscle in the Gorilla as being large and flat. In the Chimpanzee, the anterior fasciculus, according to Vrolik, is relatively stronger than in Man, and is less separated from its fellow of the opposite side. Thereby, he observes, a tendency is shown to the condition which exists in the other Apes, "in which the anterior fasciculi of the digastric are so developed as to fill up all the interspace between the two rami of the lower jaw." This expansion of the anterior portion of the muscle he has observed in the Orang and Gibbon, but it is most remarkable in the Macaques and Cynocephali. In the latter, the term "digastric" might be applied to the posterior bellies of opposite sides, for their tendons unite in front of the os hyoides to form an arch, from the convexity of which the two anterior *ventres*, which are in close proximity the one to the other, take their origin by an aponeurotic expansion. By this arrangement the power of the muscle, as a depressor of the jaw, must be considerably augmented, and it would appear to be a provision in direct relation with the more carnivorous and ferocious habits of the Baboons and Mandrills. Professor Vrolik states that in the Loris (*Stenops*) the structure of the digastricus points to a transition from the Quadrumanous type to the simple form which it takes in the Carnivora, there being only an indication of the intermediate tendon. In the Rodents (with the exception of the Rabbit), and in the Ruminants generally, it is biventral. On the contrary, in the Carnivora, the Kangaroo, the Sloths, the Elephant, the Hog, the Hyrax, it is only a single-bellied muscle. Straus-Durckheim, however, describes it in the Cat as being interrupted a little behind the angle of the jaw by a tendinous intersection, which gives rise to new fleshy fibres; and Vrolik has also noticed the occurrence of aponeurotic fibres in the

is united with the Os Occipitis. (*e*) The extent of this origin is about an inch: it is fleshy upon its outer part, viz. that from the Mastoid Process, and tendinous on the inside from the ridge. From its origin it passes forwards, downwards, and a little inwards, much in the direction of the posterior edge of the Mammillary Process; and forms a round tendon first in its center and upper surface. This tendon passes on in the same direction; and when got near the Os Hyoides, it commonly perforates the anterior end of the Stylo-Hyoideus Muscle; and from the lower edge of this tendon, some fibres seem to go off, which degenerate into a kind of Fascia, that binds it to the Os Hyoides; and some of it goes across the lower part of the Mylo-Hyoidæus, and joins its fellow on the opposite side; binding the Os Hyoides by a kind of belt. At this part the tendon

---

middle of the muscle in other Carnivora. It would almost seem as if the unity of the muscle in the Carnivores was produced by the reunion of the two fasciculi which are common to Man and the Apes. The straight line pursued by the simple muscle in these animals, as compared with its deflected course in Man, is evidently conducive to more powerful and direct action in the abduction of the maxilla. According to Cuvier, the digastric is absent in the Ant-eaters and Armadillos, its place being supplied by a long and slender muscle which takes its origin from the sternum, and is inserted into the inferior border of the jaw.] (1)

(*e*) [It is observed by Tyson that in the Chimpanzee the digastric does not arise from the mastoid process of the temporal, as in Man, but from the occipital bone. He adds, "Drelincourt describes it (the digastricus) in Apes thus: Tendinem habet intermedium pollice longum, et gracilem, enascitur antem non ab apophyse Styloide, sed ab osse Basilari." According to Vrolik, however, in the Chimpanzee it is attached to the tuberosity which represents the mastoid apophysis. In the Gorilla, Duvernoy traces its attachment to a groove behind the mastoid apophysis.] (2)

---

(1) *Vide* Cuvier, op. cit., p. 91; Duvernoy, Archives du Muséum d'Histoire Naturelle, Tome viii., p. 182; Vrolik, Recherches d'Anatomie Comparée sur le Chimpansé, Amsterdam, 1841, pp. 17, 26; Article Quadrumana, Todd's Cyclopædia of Anatomy; Straus-Durckheim, op. cit., p. 218.

(2) Orang Outang, sive Homo Sylvestris, or the Anatomy of a Pygmie, by E. Tyson, M.D., F.R.S., p. 86; Vrolik; Duvernoy, op. et loc. cit.

becomes a little broader, makes a turn upwards, inwards, and forwards, and gives origin to the anterior belly, which passes on in the same direction, to the lower part of the Chin, where it is inserted tendinous and fleshy, into a slight depression on the under, and a little on the posterior part of the Lower-Jaw, almost contiguous to its fellow. (*f*) Besides the attachment of the middle tendon to the Os Hyoides, there is a ligamentous binding, which serves, in some measure, as a pulley. (*g*) This is more marked in some subjects than in others; and this depends on the strength of the tendinous expansion, which binds the tendon of the Digastricus to the Os Hyoides.

When we say that these parts are attached to the Os Hyoides, we do not mean that they can be traced quite into it, like some other tendons in the body; but the Os Hyoides seems to be the most fixt point of attachment. Very often we find two anterior bellies to each Muscle; the uncommon one, which is the smallest, does not pass to the Chin, but joins with a similar portion of the other side, in a middle tendon, which is often fixed to the Os Hyoides. At other times, we find such a portion on one side only; in which case it is commonly fixed to the middle tendon of the Mylo-Hyoidæus.

The use of these Muscles with regard to the Lower-Jaw, is

(*f*) [The surface for insertion of the digastric varies in position in Mammalia. In Carnivora it is attached for a varying distance to the inferior border of the horizontal ramus. In the Cat it is inserted, according to Cuvier, into the angle of the jaw. Straus-Durckheim, however, describes it as reaching the anterior half of the lower border. In the Rodents it is prolonged to the back part of the arch of the chin, into which it is fixed; in the Elephant it is attached to the posterior border of the maxilla; in the Horse the principal portion of the muscle is inserted into the angle; in the Ruminants it extends as far as the mid-length of the ramus.] (1)

(*g*) [The existence of an apparatus analogous to the pulley was denied by Monro primus (*vide infra*, note *h*). Many writers describe the tendon of the digastricus as passing through an aponeurotic sheath lined by a synovial membrane. Such a structure, although in some instances apparent enough, is by no means constant.

---

(1) Cuvier; Straus-Durckheim, op. cit.

principally to depress it; but according as one acts a little more forcibly than the other, it thereby gives the Jaw a small rotation; and becomes, in that respect, a kind of antagonist to the Pterygoidæus Externus. Besides depressing the Lower-Jaw, when we examine the dead body, they would appear to raise the Larynx. But although they have this effect, a proper attention to what happens in the living body, will probably shew, that their principal action is to depress the Lower-Jaw, and that they are the Muscles which are commonly employed for this purpose. Let a Finger be placed on the upper part of the Sterno-Mastoidæus Muscle, just behind the posterior edge of the Mastoid Process, about its middle, touching that edge a little with the finger; then depress the Lower-Jaw, and the posterior head of the Digastric will be felt to swell very considerably, and so as to point out the direction of the Muscle. In this there can be no deception; for there is no other Muscle in this part that has the same direction; and those who are of opinion that the Digastric does not depress the Lower-Jaw, will more readily allow this, when they are told, that we find the same head of the Muscle act in deglutition; but not with a force equal to that which it exerts in depressing the Lower-Jaw. (*h*) Further, if the Sterno-Hyoidei, Sterno-Thyroidei, and Costo-Hyoidei, acting at the same time with the Mylo-Hyoidei, and Genio-Hyoidei, assisted in depressing the Jaw, the Os Hyoides, and Thyroide Cartilage, would probably be depressed, as the

---

(*h*) [Hunter here probably refers to a paper by Alexander Monro primus on the Articulation Muscles and Luxation of the lower jaw, which appeared in the first volume of the Edinburgh Medical Essays. Monro opposes the commonly-received opinion as to the function of the digastric by the following arguments: That the shortness of the bellies of the digastric does not admit so large a contraction as is often required; —that "the proportional force of the digastric to that of the elevators of the jaw is considerably less than is seen in other parts of the body where antagonist muscles are, which proportional force of these muscles is on some occasions greatly lessened by the angle of insertion of these digastric muscles into the jaw, decreasing as the mouth is opened." He denies the existence of any mechanism akin to a pulley, such as is seen in the larger oblique muscle of the eye. He describes the connection of the digastric with the os hyoides as taking place simply by a firm

bellies of the Sterno-Hyoidei, and of the other lower muscles, are by much the longest; but on the contrary, we find that the Os Hyoides, with the Thyroide Cartilage, is a little raised in the depression of the Jaw, which we may suppose to be done by the anterior belly of the Digastric: (*i*) and secondly, if these

---

aponeurosis, which comes off from the round tendon of the muscle, and is in part united to the os hyoides and in part joins the aponeurosis of the opposite side. He denies that there is anything like a sheath in which the middle tendon slides. Lastly, he describes some experiments on the living and dead subject, which he believes disprove the agency, at least, to any great extent, of the anterior or posterior belly of the digastricus in depressing the jaw. He believes the action of the muscle to be principally exercised in deglutition, in which act it pulls the os hyoides upwards, so as to press the root of the tongue against the soft palate. The real depressors of the maxilla he considers to be the sterno-hyoidei and genio-hyoidei, together with the coraco-hyoidei and mylo-hyoidei, which may be looked upon as digastrics whose middle intersection is at the hyoid bone; and also the sterno-thyroidei, thyro-hyoidei, hyo-glossi, and genio-glossi, which, in their united action, are to be considered as two many-bellied muscles acting on the lower jaw. (1)

The opinion which is now generally held, and which is probably correct, may be taken as a mean between the opposite views of Hunter and Monro. Anatomists generally allow that the principal instrument in the depression of the jaw is the digastric; but in order to the production of that effect it is necessary that the os hyoides should be fixed, which is done by the agency of the muscles that pass between that bone and the sternum and scapula, so that the latter become indirectly depressors of the maxilla. It is also generally admitted that all the other muscles which pass from the os hyoides to the maxilla when the former is fixed, may be accessories of the digastric in this act. When, on the contrary, the jaw is fixed, the contraction of the digastric must raise the os hyoides and the parts attached to it. This action of the digastric is called into play in deglutition when, together with the stylo-hyoid, mylo-hyoid, and genio-hyoid, it elevates the hyoid bone, and with it the base of the tongue. In swallowing the mouth is first closed.]

(*i*) [Monro states that the os hyoides moves downwards and forwards when the mouth is opened. In observations on myself and others I have never been able to arrive at the conviction that the hyoid bone and thyroid cartilage are raised when the lower jaw is depressed, as is asserted in the text.]

---

(1) Medical Essays and Observations by a Society in Edinburgh, vol. i., 1733.

Muscles were to act to bring about this motion of the Jaw, these parts would be brought forwards, nearer to the straight line between the Chin and Sternum, which is not the case in this action; whereas we find it to be the case in deglutition, in which these evidently act. By applying our fingers upon the Genio-Hyoideus, and Mylo-Hyoideus, near the Os Hyoides, between the two anterior bellies of the Digastric, (not near the Chin, where the action of these two bellies may occasion a mistake), we find these Muscles quite flaccid; which is not the case in deglutition, nor in speaking, in which they certainly do act; nor do we find the Muscles under the Os Hyoides at all affected, as they are in the motion of the Larynx.

It has been observed, that when we open the mouth, while we keep the Lower-Jaw fixed, the fore-part of the Head or Face is necessarily raised. Authors have been at a good deal of pains to explain this. Some of them considered the Condyles of the Jaw, as the center of motion; but if this were the case, that part of the Head, where it articulates with the Spine, and of consequence the whole body, must be depressed, in proportion as the Upper-Jaw is raised; which is not true in fact. Others have considered the Condyles of the Occiput as the center of motion; and they have conceived the Extensor Muscles of the Head to be the moving powers. The Muscles which move the Head in this case, are pointed out by two circumstances, which attend all muscular motion; in the first place, all actions of our body have Muscles immediately adapted to them; and secondly, when the mind wills any particular action, its power is applied by instinct to those Muscles only, which are naturally adapted to that motion; and further, the mind being accustomed to see the part move which is naturally the most moveable, attends to its motion in the volition, although it be in that instance fixed, and the other parts of the body move towards it; and although the other parts of the body might be brought towards it by other Muscles, and would be so, if the mind intended that they should come towards it, yet these Muscles are not brought into action. Thus the Flexors of the Arm commonly move the hand to the body; but if the hand be fixed, the body is moved by the same Muscles to the hand. In this case, however, the mind wills the motion of the hand towards the body, and brings the Flexors into action; whereas if it wished to bring the body towards the hand, the Muscles of the forepart of the body would be put into action, and this would produce the same effect.

To apply this to the Lower-Jaw; when we attempt to open the Mouth, while the Lower-Jaw is immoveable, we fix our attention upon the very same Muscles (whatever they are) which we call into action, when we depress the Lower-Jaw; and we find that we act with the very same Muscles; for our mind attends to the depressing of the Jaw, and not the raising of the face; and under such circumstances the mouth is actually opened. We find then by these means the head is raised; and the idea that we have of this motion, is the same that we have in the common depression of the Jaw; and we should not know, except from circumstances, that the Jaw was not really depressed; and we find at this time too, that the Extensors of the head are not in action. On the contrary, when the Jaw is fixed in the same situation, if we have a mind to raise the head, or Upper-Jaw, which of course must open the mouth, we fix our attention to the Muscles that move the head backwards, without having the idea of opening our mouth; and at this time the Extensors of the head act. This plainly shows, that the same Muscles which depress the Jaw, when moveable, must raise the head, when the Jaw is kept fixed.

This is a proof too, that there are no other Muscles employed in depressing the Lower-Jaw, than what will raise the head under the circumstances mentioned. This will further appear from the structure of the parts; wherein four things are to be considered, viz. the articulation of the Jaw; the articulation of the Head with the Neck; the origin, and the insertion of Digastric Muscle.

Suppose A, the Upper-Jaw, to be fixed, and the Lower-Jaw B, to be moveable on the Condyle C: if the Digastric contracts, its origin E, and insertion F will approach towards one another; in which case it is evident, that the Lower-Jaw will move downwards and backwards. But if the Lower-Jaw be fixed, as in the case supposed, and the Vertebræ G G G be also fixed, the Condyle will move upwards and forwards upon the eminence in the joint, the fore-part of the head will be pushed upwards and backwards by the Condyle, and the hind-part of the head will be drawn down; so that the whole shall make a kind of circular motion upon the upper Vertebræ; and the Digastric Muscle pulling the hind-part of the head towards the Lower Jaw, and at the same time pushing up the Condyles against the fore-part of the head, acquires, by this mechanism, a very considerable additional power.

OF THE STRUCTURE OF A TOOTH. AND, FIRST, OF THE ENAMEL.

A Tooth is composed of two substances, viz. Enamel and Bone. (*l*) The Enamel, called likewise the vitreous, or cortical

---

(*l*) [It may be stated, in general terms, that a tooth is a hard body primarily distinct from the skeleton, situated in the mouth or at the commencement of the alimentary canal, composed of organic and inorganic matter, and enclosing a vascular pulp which occupies a cavity in its interior. In examining the teeth of the greater number of vertebrate animals, we are able to discriminate two or more tissues, of differing structure and hardness. The principal dental tissues are the *dentine* or *ivory* (bone of the text), the *enamel*, and the *cement* or *crusta petrosa*. Of these, the dentine forms the chief mass of the body of the tooth, the cement the outer crust, whilst the enamel, where it exists, is external to the dentine, or between it and the cement. Besides the three principal dental tissues, various modifications of the dentine not unfrequently occur, to which names have been applied indicative of their structural peculiarities—*e.g.*, *osteo-dentine*, *vaso-dentine*. Although structures analogous to dental organs are met with in the invertebrate classes, it is amongst the vertebrata alone that true calcified teeth are found. In the teeth of vertebrate animals, the number of component tissues, and the manner of their arrangement, exhibit considerable variation. Amongst the class *Pisces*, there are a few instances (*e.g.*, the Wrasse *Labrus*) in which the teeth consist of but one tissue, which is always a modification of dentine. It is more common, in the same class, to find the tooth

part, is found only upon the body of the Tooth, and is there composed of ordinary dentine, and of the same structure, permeated by vascular canals (vaso-dentine): in such instances, the hard unvascular dentine is external, and supplies the place of the enamel of higher animals. The molars of the Dugong afford examples of teeth composed of dentine and cement alone, the latter forming a thick crust on the surface of the former. At first sight, the human teeth and those of Carnivora might be supposed to consist of but two tissues—dentine and enamel. This is the statement of the text, and until a comparatively late period it passed as an acknowledged fact. It was on account of this apparent simplicity of structure that these teeth were distinguished as "simple" by the Cuviers. More recent observation, however, has shown that, besides the enamel and dentine, the third principal dental tissue—the cement—forms a component part of the human and the other so-called simple teeth. The teeth of Mammalia, which received from Cuvier the name of "Compound" or "Complex," differ from those of Man and the Carnivora only in a different proportion and disposition of the same constituent tissues. In some instances—*e.g.*, the molars of the Elephant, the molars of the African Wart-hog (*Phacochœrus*), the pectinated incisors of the Flying Lemur (*Galeopithecus*)—a tooth of a high degree of complexity is produced by the aggregation of distinct denticles. The great grinding tooth of the Elephant is composed of a series of denticles, each having the form of a plate, vertical to the grinding service, and transverse to the long diameter of the tooth. Each plate consists of dentine coated by an investment of enamel, whilst the interspaces between the denticles are completely filled by cement. In this fully-formed tooth, the bases of the several divisions of the crown become fused into a common body of dentine; but at an earlier period, before the calcification of the plates is complete, they are held together by the cement alone. The incisors of Galeopithecus, on the other hand, differ from the tooth just described in the circumstance that the denticles project as distinct processes from the base of the crown: each consists of dentine with a covering of enamel, which is again coated with a layer of cement of extreme thinness; at the base the denticles are united, and into each a prolongation is continued from the common pulp cavity. But a high degree of complexity of dental structure is not restricted to Mammalia. In the grinding apparatus of the Parrot-fishes (*Scarus*), for instance, which "browse upon the lythophitas that clothe as with a richly-tinted carpet the bottom of the sea, just as the Ruminant Quadrupeds crop the herbage of the dry land," (1) dental masses, bearing a close analogy to the complex molars of the Elephant, are formed by an aggregation of denticles. Moreover, each denticle of the Parrot-fish consists of dentine with a thick covering of enamel: the denticles are united

---

(1) Owen, Art. Teeth, Todd's Cyclop. Anat. and Phys.

laid all around, on the outside of the bony, or internal substance.

---

together by a partial filling of their interspaces by cement, whilst a fourth substance is added, in the form of a vascular osteo-dentine, which, produced by the ossification of the base of the pulp, serves to unite the denticles with each other, and with the subjacent pharyngeal bone. The tooth, thus composed of four substances, exhibits the highest degree of complexity yet observed in the animal kingdom. Many other variations in the structure of the teeth, and in the arrangement of the dental tissues, are to be found amongst vertebrate animals; but sufficient examples for the purpose of illustration have been adduced. (1)

In the human tooth, the dentine forms the principal mass both of the fang and crown: in the latter situation it is covered by enamel, in the former by cement. As no separate mention is made of this last-named substance in the text, it will be, perhaps, convenient to describe its disposition and structure in the human tooth before commencing our commentary on the enamel.

The *Cement*, *Crusta Petrosa*, *Substantia Ostoidea*, is a layer of true bone which in Man covers the fang or fangs of the tooth. In some instances it appears to commence where the enamel ceases, but in many others it manifestly overlaps the enamel for a short distance from its edge. The layer of cement is extremely thin at the neck of the tooth; but, in tracing it down the fang, it is found to become thicker as it descends, until it attains its maximum of depth at the apex. That confluent condition of the fangs which is so common in certain of the molar teeth amongst the leucous and xanthous varieties of Man, is produced by the cement which fills the interspaces, just as in the compound molar of the Elephant it unites a series of denticles into a single mass. It is, of course, found of considerable thickness along the sulci which indicate the division of fangs so enjoined. According to some of the best observers, a thin layer of cement may be traced in the unworn tooth covering the enamelled crown. In a paper read before the Medico-Chirurgical Society in 1839, Mr. Nasmyth announced the discovery of a membrane separable by the action of dilute hydrochloric acid from the surface of the enamel, which, he expressly states, is continuous with the cement, and which, he believed, was at least analogous to, if not identical with, a layer of that substance in a rudimentary condition: this layer he terms the *persistent capsular investment*. (2) Nasmyth's observation was confirmed by Professor Owen, who, in writing of the so-called simple teeth, " says their crowns are originally and their fangs are always covered by a thin coat of

---

(1) Owen, op. cit., Odontography.
(2) Med.-Chir. Trans., vol. xxii., pp. 312-315. Researches on the Development, Structure, and Diseases of the Teeth, by Alex. Nasmyth, p. 79; 1849.

It is by far the hardest part of our body; insomuch that the

---

cement." (1) Other observers, amongst whom are Huxley and Kölliker, although allowing the existence of a thin calcified pellicle (by the former named '*Nasmyth's membrane*,' by the latter the '*cuticle of the enamel*') on the surface of the enamel, do not recognise in it a prolongation of the cement. It would appear that the layer in question is remarkable for its great capability of resisting chemical reagents; it differs in this respect from ordinary cement, and is proportionately better fitted to protect the crown of the tooth. The original surmise of Nasmyth has, however, received some confirmation from certain observations recorded by Mr Tomes in his 'System of Dental Surgery.' This gentleman, in describing the disposition of the cement, writes: " In a few rare instances it may be traced, not only over the terminal edge of the enamel, but for some little distance upon the coronal portion of the tooth ; and specimens are now and then found in which it fills up the deep fissures situated between the tubercles of the molar teeth." Again, in reference to Nasmyth's membrane, he says : " In several specimens which have been decalcified after being reduced sufficiently thin for microscopic examination, this membrane is obviously continuous with the cementum of the fang ; and in other specimens which have not been treated with acid, I find the membrane thickened in the deep depressions of the crowns of molar teeth, and there tenanted by a distinct lacuna." (2)

As life advances, the cement increases in thickness, especially towards the apex of the fang. In those cases in which its formation extends beyond a certain limit, it constitutes the disease known as "dental exostosis." The inner surface of the cement is so intimately connected with the dentine, that under a high degree of magnifying power its boundary line is not very sharply defined. Its external surface, which is uneven and not unfrequently marked with circular striæ, is in close contact with the alveolar periosteum and the gum ; but it is more firmly united with the former than with the latter. Its tissue is less hard than either dentine or enamel : consequently, its surface when exposed to attrition becomes worn down at an earlier period, as may be seen in the complex grinders of the Elephant, the Masked Boar, or Capybara. In chemical composition the cement is almost identical with bone. The following is its analysis according to Bibra : (3)

	In Man.	In the Ox.
Organic substance	29·42	32·24
Inorganic substance	70·58	67·76
	100·00	100·00

---

(1) Article Teeth, Todd's Cyclop., p. 867.
(2) Tomes, ' A System of Dental Surgery,' pp. 257, 271 ; 1859.
(3) Manual of Human Microscopic Anatomy, by A. Kölliker, p. 297 ; 1860.

hardest and sharpest saw will scarce make an impression upon

In the Ox the proportions of the several ingredients were—	
Phosphate of Lime and Fluoride of Calcium	58·73
Carbonate of Lime	7·22
Phosphate of Magnesia	0·99
Salts	0·82
Cartilage	31·31
Fat	0·93
	100·00

The cement may be readily deprived of its earthy salts by the action of acids. The organic residue yields ordinary gelatine on boiling.

Cement, we have already said, is a layer of true bone, with but little modification. But it is only in situations in which it attains great thickness (e.g., the teeth of the Horse, Sloth, Ruminants), that it exhibits canals for the passage of vessels (Haversian canals). In the human tooth the existence of Haversian canals is to be considered exceptional, although they are occasionally to be met with in situations where the cement is of unusual thickness. When, for instance, the fangs of a molar are united by cement, a vascular canal is not unfrequently found traversing the medium of union (Tomes.) Their existence, therefore, in the human tooth is not to be regarded as necessarily an indication of morbid action, although they are most frequently to be noticed in the thickened cement of old teeth and in cases of dental exostosis. When present, they penetrate from without; in some instances they branch once or twice, and anastomose, or end by blind extremities, or a canal may end by a dilatation, "as though occupied by a vessel that had turned upon itself and gone out by the same channel through which it entered." (1) Their diameter is too narrow (0·003''' to 0·01''', Kölliker) to admit the presence of marrow as well as of blood-vessels. As in bone, the vascular canal in cement is surrounded by concentric laminæ.

Cement, like other osseous tissues, is composed of a matrix, in which are imbedded more or less numerous microscopic cells (*corpuscles* or *lacunæ*), from which proceed, in various directions, fine, delicate ray-like canals (*canaliculi*), which by anastomosing constitute a system of tubes for the convection of nutrient fluid through the tissue. The *matrix* or *basal substance* usually appears to consist of lamellæ, concentrically placed, the centre of the tooth being the common centre. According to Kölliker, the basal substance is sometimes granular, sometimes striated in the transverse direction, sometimes more amorphous, frequently lamellated. Mr Tomes describes it as granular, and likens it to a mass of coherent fig-seeds or to oolite. The cells or *lacunæ* are scattered through the matrix with some degree of regularity, being frequently,

---

(1) Tomes, Lectures on Dental Physiology and Surgery, p. 58 ; 1848.

it, and we are obliged to use a file in dividing or cutting it. (*m*)

although not always, disposed in series; or they appear as if placed between concentric laminæ. They are always absent near the neck of the tooth, where the cement is thinnest; they generally appear about the middle of the fang, and increase in number with the thickness of the cement towards the apex, where they often present the serial appearance to which we have just referred. They possess the essential characters of the lacunæ of bone. They appear dark by transmitted, and white by reflected light; a circumstance due simply to their being filled with air. In the lacunæ of fresh osseous tissue, a delicate-walled cell, containing a clear fluid and a single nucleus, has been discovered by Virchow. From this included cell, numerous processes are sent into the canaliculi which branch out on every side from the lacuna. These processes anastomose with like processes from the included cells of neighbouring lacunæ. In shape the lacuna is generally oval, but occasionally round or fusiform; in fact, in this respect they are subject to great variety. Equally variable is their size; according to Kölliker, $0.005'''$ to $0.02'''$, even to $0.03'''$, in diameter. In the thick layers of cement which are found on old teeth, these recesses are generally of an elongated form and very irregular. Some lacunæ, either separately or in groups, may be seen to be surrounded by a tissue of more than the ordinary transparency. This clearer substance is well defined by a slightly-waved border, which Kölliker supposes may be in some way related to the cell from which the lacuna is formed. The *canaliculi* which are given off by the lacunæ of cement are of unusual number and length. They are frequently so numerous as to give the idea of a pencil or plume of processes, which again branch into still finer rays. The majority of the canaliculi are directed outwards towards the surface of the tooth, or inwards towards the dentine: some, however, follow the course of the length of the tooth; or, where a vascular canal is present, they take a direction towards its surface. In crossing the matrix, they give it a finely-striated appearance; they anastomose freely with the canaliculi of neighbouring lacunæ, and near the surface of the dentine they may be seen to inosculate with the terminations of the dentinal tubes. Other cavities may be observed in cement; some of them are of a pathological nature. But, besides these, Mr Tomes notices tubes of occasional occurrence which pass across the cement towards the surface of the tooth. They equal in size the dentinal tubes, but have fewer branches. Occasionally the canaliculi enter them. Kölliker, in referring to what appear to be the same tubuli, says that their branches are frequently connected with the terminations of the dentinal canals, and with the canaliculi. On the fangs of deciduous teeth the cement is thin, and exhibits comparatively few lacunæ.] (1)

(*m*) [Such is the hardness of the enamel, that it emits sparks when

---

(1) Kölliker, Tomes, Owen, op. cit.

When it is broken it appears fibrous or striated; and all the fibres or striæ are directed from the circumference to the center of the Tooth.

This, in some measure, both prevents it from breaking in mastication, as the fibres are disposed in arches, and keeps the Tooth from wearing down, as the ends of the fibres are always acting on the food. (*n*)

The Enamel is thickest on the grinding surface, and on the cutting edges, or points, of the Teeth; and becomes gradually

---

struck with steel. (1) It is also more brittle than the cement and dentine. We have already referred to the worn surface of the Elephant's molar, as well exemplifying the relative density of the dental tissues. It presents a series of transverse cylindrical ridges of enamel, each enclosing a depressed surface of dentine, whilst the plates so formed are separated from each other by still more deeply depressed valleys of cement.]

(*n*) [The external surface of the enamel appears smooth, but almost always presents, on careful examination, delicate, closely-approximated transverse ridges, and in some instances annular elevations. We have already adverted to the fact that it is covered by a delicate calcified pellicle (Nasmyth's membrane, cuticle of the enamel). According to Kölliker, " Nasmyth's membrane " is a calcified structureless membrane, 0·0004″′ to 0·0008″′ in thickness, which on its internal surface is furnished with depressions for the reception of the extremities of the enamel fibres. Its connection with the enamel is so intimate, that it can only be demonstrated by the employment of hydrochloric acid. Professor Huxley believes that the membrane in question is the altered *membrana preformativa* of the dental pulp. To this subject we shall have, however, to recur, when stating the various opinions current on the subject of the development of the dental tissues.

In very thin sections, enamel is translucent and of a bluish colour. Its substance is generally allowed to be composed of the so-called enamel fibres (enamel-prisms, enamel-needles)—hard, dense, microscopic structures, consisting almost entirely of earthy matter. These fibres were described by Retzius as being solid hexagonal prisms. They are not, however, regular in outline. Mr Tomes observes, that in a transverse section he has as frequently found them approach a square or an irregular circle as any other form. They are long, extending for the most part throughout the thickness of the enamel, the inner end resting on the dentine, the outer being in contact with Nasmyth's membrane. The position of the enamel fibres being generally vertical to the surface

---

(1) Nasmyth's Researches, p. 81.

thinner on the sides, as it approaches the neck, where it termi-

of dentine on which they rest, it follows that their direction will be horizontal on the sides of the tooth, vertical on the summit of the crown. Fibres occupying all the intermediate positions between the vertical and horizontal will be seen in examining the enamel as it passes from the grinding or cutting surface downwards to the side of the tooth. We have said that the fibres, for the most part, extend throughout the entire breadth of the enamel, and this is undoubtedly true of the greater number; but, as the surface of the dentine is more or less conical, and no separation of the fibres takes place as they proceed outwards, neither has any increase in their thickness been observed, we cannot doubt the existence of shorter supplemental fibres, which fill up the intervals that otherwise would necessarily occur. Such supplemental fibres, however, are not easily demonstrated, owing to the waved course of the fibres, and the consequent difficulty in tracing a series of them throughout their entire length. It is almost impossible to isolate for any distance the enamel fibres in the adult tooth, although they can be readily seen in transverse and longitudinal sections : it may, however, be more readily accomplished in the softer enamel of young or developing teeth. The breadth of the enamel fibre is, according to Kölliker, $0.0015'''$ to $0.0022'''$. Isolated prisms obtained from young enamel are seen, especially after treatment with dilute hydrochloric acid, to have a slightly varicose outline, and to be marked with transverse lines, or striæ, which occur at tolerably regular intervals, and produce an appearance similar to that of a "colossal muscular fibrilla." (1) This striated aspect led Mr Nasmyth to the opinion that enamel consisted of "nothing more than a mass of cells arranged in rows, and fitted closely together, but held only slightly in contact by a thin web of membrane." (2) Retzius and others have also believed the transverse marks to be the indications of pre-existing walls of coalesced cells. Kölliker, on the other hand, maintains that they "are the expression of the growth of the fibres through apposition, or are *not* the expression of their composition of cellules." (3) Mr Tomes observes that the transverse striæ are much more strongly marked in some specimens than in others, and that they are most evident in those portions of enamel which, when seen by transmitted light, are of a brown colour. This condition will sometimes be found in certain unhealthy subjects to pervade the entire enamel. In such cases he has found the markedly striated appearance to result from the occurrence, at regular intervals, of minute granular masses in the central portion of the enamel fibres. These masses are comparatively opaque, and the alternation of opaque and transparent parts gives the appearance of striation. This appearance is rendered more distinct by the use of

---

(1) Kölliker, op. cit., p. 205.
(2) Nasmyth, op. cit., p. 83.
(3) Kölliker, op. et loc. cit.

nates insensibly, though not equally low, on all sides of the

dilute hydrochloric acid, which has the effect of removing the granular masses from that portion of the tissue with which it comes more directly in contact, whilst in deeper parts the opaque masses become more distinct in consequence of the removal of all opacity from the superficial portion. A transverse section thus treated with dilute hydrochloric acid has the appearance of a portion of honeycomb from which the honey has been removed. As the result of his observations, Mr Tomes does not admit that the structure of fully-formed enamel is in the strict sense of the term "fibrous." "A honeycomb," says this observer, "if the cells were filled with a material of greater opacity or density than the wax of which the cells themselves are formed, would not be regarded as fibrous ; yet the arrangement of the parts would resemble those of the enamel." To the argument that the cleavage of the enamel demonstrates its fibrous nature, he replies that the lines of fracture do not run in the longitudinal interspaces between the fibres, but through the lines of granular masses. He, however, allows that in the young tooth, during the process of formation, the fibrous arrangement is sufficiently distinct, and may be demonstrated by hydrochloric acid. In the most perfectly-developed enamel, he finds the sheaths of the fibres completely blended ; the longitudinal and transverse markings are comparatively faint, and appear under a high magnifying power, with a good light, not as dark, but as light lines, enclosing spaces occupied by a more dense and opaque material. He describes several imperfect forms of development in this tissue, all of which are to be regarded as predisposing causes of caries : the fibrous condition may be maintained in consequence of the imperfect blending or fusion of the sheaths of the fibres ; or the central portion or contents of the fibre may be imperfectly developed, remaining in the condition of fine globular masses or granules ; or fine cavities arranged in single lines may occupy the centre of the fibre—in some cases these may coalesce, and convert the fibre into a tube ; or, lastly, both longitudinal and transverse markings may be replaced by a general granular condition of the tissue. (1)

There can be no doubt that the conjunction of the enamel fibres is very intimate. Neither intermediate substance nor regular canals have been demonstrated between them. Cavities, however, frequently exist in enamel, and they are referrible either to extension of the dentinal tubes and of elongated cavities produced by the enlargement of the dentinal tubes into the enamel, or to irregular fissures in the outer and middle parts of that tissue. These latter are frequently found leading down from the depressions between the cusps of the molar and premolar teeth, and are to be placed in the foremost rank amongst the predisposing causes of caries.

Like the tubes of dentine, the enamel fibres do not pursue a straight

---

(1) Tomes, System of Dental Surgery, p. 258—277.

Teeth. (*o*) On the base or grinding surface it is of a pretty equal thickness, and therefore is of the same form with the bony substance which it covers.

It would seem to be an earth united with a portion of animal substance, as it is not reducible to quick lime by fire, till it has first been dissolved in an acid. When a Tooth is put into a weak acid, the Enamel, to appearance, is not hurt; but on touching it with the fingers, it crumbles down into a white pulp. The Enamel of Teeth, exposed to any degree of heat, does not turn to lime: it contains animal mucilaginous matter; for when exposed to the fire, it becomes very brittle, cracks, grows black, and separates from the inclosed bony part of the Tooth. (*p*) It is capable, however, of bearing a greater degree of heat than the bony part, without becoming brittle and black.* This substance

* From this circumstance we can shew the Enamel better by burning a Tooth, as the bony part becomes black sooner than the Enamel. The method of burning, and shewing them after they are burnt, is as follows.—Let one half of a Tooth be filed away, from one end to the other, then burn it gently in the fire; after this is done, wash the filed surface with an acid, or scrape it with a knife. By this method you will clean the edge of the Enamel, which will remain white, and the bony part will be found black.

course. On the contrary, they are everywhere slightly and irregularly waved. On the summit of the crown they present two or three larger waves. Decussations of layers of the enamel prisms take place in transverse planes of the tooth. Hence it is that, on the addition of hydrochloric acid, a longitudinal section will have a striated appearance, the prisms cut transversely will appear darker than those which present their longitudinal aspect. Layers of the enamel fibres may be seen on the summit of the crown to run in an annular form, describing circles on the molar, ellipses on the cutting teeth. Besides the striated appearance indicating the stratified arrangement of the fibres, Kölliker describes certain brownish or colourless lines crossing the enamel fibres in different directions, which in perpendicular sections are seen as obliquely ascending lines or arches, but in transverse sections as circles in the outer layers of the tissue, rarely occurring throughout its entire substance. He regards them as " the expression of the lamellated mode of formation of the enamel."] (1)

(*o*) [The enamel extends lower upon the inner and outer than upon the opposed surfaces of the crowns.]

(*p*) [Enamel in chemical composition differs from the other tissues of

(1) Kölliker, op. cit., pp. 296, 297.

has no marks of being vascular, and of having a circulation of fluids: the most subtile injections we can make never reach it; it takes no tinge from feeding with madder, even in the youngest animals; and, as was observed above, when soaked in a gentle acid, there appears no *gristly* or *fleshy* part, with which the earthy part had been incorporated.*

We shall speak of the use and formation of the Enamel hereafter, when they will be better understood.

OF THE BONY PART OF A TOOTH.

The other substance of which a Tooth is composed, is bony; but

---

* In all these experiments I never could observe, that the Enamel was in the least tinged, either in the growing or formed Tooth. This looks as if the Enamel were the earth more fully depurated, or strained off from the common juices in such a manner, as not to allow the gross particles of madder to pass. Here it may not be amiss to remark, that the names given to animal substance, such as Gluten, &c. are not in the least expressive of the thing meant; for there is no such thing as glue in an animal, till it has either undergone a putrefactive process, or been changed by heat. And here too I would be understood, that I call earth no part of an animal; nor does it make up any part of an animal substance.

---

the tooth in the exceedingly small proportion of animal matter which it contains. The organic matter differs from gelatine, and is in every respect analogous to the substance of the epithelium. The following is Bibra's analysis: (1)

	Molar Tooth in an Adult Man.
Phosphate of Lime, with some Fluoride of Calcium	89·82
Carbonate of Lime	4·37
Phosphate of Magnesia	1·34
Salts	0·88
Organic substance	3·39
Fat	0·20
	100·00
Organic substance	3·59
Inorganic parts	96·41

---

(1) Kölliker, op. cit., p. 205.

much harder than the most compact part of bones in general. (q)

(q) [In the following description of the minute structure of dentine (termed, in the text, the bony part of a tooth), the account given by Kölliker has been principally although not solely followed:

The *dentine, ivory, substantia eburnea, ebur*, is of a yellowish white colour, and appears in the fresh specimen to a certain extent transparent or translucent. In the dried tooth it is white, and has a silky lustre, from the circumstance that it is permeated in every part of its substance by a series of microscopic tubes which, in that condition, contain air. It consists of a matrix and of the tubes just mentioned, *dentinal tubules, dental canals, canaliculi dentium*. The dentine alone bounds the pulp cavity, with the exception of a small portion at the extremity of the root; and in the uninjured tooth it is never exposed, being covered everywhere by enamel or cement. The matrix is perfectly homogeneous, showing, according to Kölliker, no trace of organised structure. The dentinal canals measure in width 0·0006''' to 0·001''', some of those at the root 0·002''': they commence by open mouths at the pulp cavity, and proceed outwards throughout the entire thickness of the tissue. Each canal appears to have a special wall in the shape of a yellowish ring, which is generally visible when the tubes are transversely divided. Kölliker believes, however, that the yellowish ring surrounding the tube is not to be considered as the real wall. "The apparent walls," he writes, " of the dentinal tubes, which are usually seen upon the transverse sections, are not the true walls of the canals, but rings, the appearance of which arises from this, that a certain length of the canals is always seen with the microscope in the thickness of the section, never sufficiently fine to obviate this effect; and the short tubular segments being curved in direction, a greater apparent thickness is thus given to the walls than they really possess. If, upon a transverse section, the openings of the canals be brought accurately into focus, we perceive, instead of a dark ring, only a yellowish, very narrow edge; and it is this that I regard as being the true wall." (1) During life, the canals are generally believed to be permeated by a clear, pellucid fluid; in the dried state, they become filled with air, and appear as dark lines by transmitted, and shining filaments by reflected light. The dentinal tubes undulate in their course; they present two or three larger flexures, and a great number of smaller ones—according to Retzius as many as 200 in a line. As they proceed outwards, they divide, branch, and anastomose. The divisions, which are frequent near the origin of the tubes, are generally dichotomous; they may be repeated from two to five times or more—so that one tube may form four, eight, sixteen, or more tubes. Their calibre becomes diminished by division; and they run in a nearly parallel

---

(1) Kölliker, op. cit., p. 292.

This substance makes the interior part of the body, the neck,

manner, and close together towards the surface of the dentine. In the crown of the tooth, the dental tubes, at about the middle or outer third of their course, commence sending off fine ramifications, which are mostly simple, but sometimes branched. In this situation, a tube will frequently appear to terminate by dividing into two fine branches. In the fang, the ramifications are much more numerous than in the crown; they commence earlier in the course of the tube, and sometimes give it a plumose, or, when the ramifications are long and branched, a brush-like appearance. These ramifications, by anastomosing, serve to connect with each other neighbouring or more remote canals. The terminations of the dentinal tubes are fine in proportion to their amount of ramification; they frequently appear as extremely delicate pale lines, like the fibrils of connective tissue, and become at length so attenuated that they cannot be farther traced. Where, however, the tubes and their ramifications can be followed to their termination, they are either found to form loops in the substance of the dentine itself by the junction of one tubule with another (*terminal loops of the dental canals*), or to end in a granular layer to be hereafter noticed;—or, passing across the boundary of the dentine, they may be traced into the enamel or crusta petrosa. (1)

Dentine presents indications of lamellation. In a longitudinal section of a tooth, arched lines more or less parallel to the circumference of the crown, and situated at a varying distance from each other, are to be observed. In transverse sections they appear as rings. These are the *contour lines* of Professor Owen; so called from their general similarity with the contour of the tooth. Mr Salter, who has described their appearance and arrangement with great minuteness, states, however, that the course of the contour line never exactly corresponds with the circumference of the crown of the tooth. "The contour of the two," he writes, "is not identical, for the markings (in whatever part examined) are more divergent than the outline of the tooth, and, passing from within outwards, abut in succession upon the external surface of the dentine. In comparing the absolute contour of any tooth, it will be found that the angle formed by its sides is more acute at its summit, or the summit of any particular cusp, than the contour markings within." (2) In teeth which have more than one cusp, the upper contour markings are confined to their own cusps, and their extremities abut against their sides; the lower ones are continuous with those of continuous cusps. These contour lines or markings are curved in their course: according to the observer just quoted, the curves are "in pro-

---

(1) *Vide* Kölliker, op. cit., p. 291.
(2) On certain Appearances occurring in Dentine, by S. J. A. Salter, Quart. Journ. Micros. Sci., vol. i., p. 254.

and the whole of the root of a Tooth. It is a mixture of two

portion to the primary curves of the dentinal tubes at any particular spot, and cross them at right angles." (1) These lines vary in human teeth in intensity and in number. In the teeth of some animals they present a beautiful appearance; especially is this the case in some of the Cetacea and Pachydermata, and in the Walrus. A tendency to break into lamellæ is not unfrequently found in these and in fossil teeth, and is sometimes indicated in fresh human teeth and in tooth cartilage. In the human tooth, the extremities of the better defined contour lines terminate in irregularly-shaped cavities (*interglobular spaces*, Czermak), situated at the surface of the dentine, immediately within the cement and enamel. Mr Salter describes these spaces or cavities as being more or less club-shaped, with the butt end of the club towards the surface, and the pointed or attenuated end stretching obliquely inwards and upwards towards the pulp cavity. They vary in size; their walls are formed by spherical masses or globules of dentine, which project into the cavity; and they are traversed by the dentinal canals, the number of which will vary with the size of the cavity. The dental canals permeate the dentine globules, five or six traversing a large globule: in the dried specimen, when the cavity contains air, the individual tubules may be traced from one globule to another, skipping, so to speak, the interglobular space. Kölliker asserts that, during life, the spaces are filled with a soft substance resembling tooth cartilage, which is permeated by the tubules. He says that this soft substance "offers more resistance to hydrochloric acid than the matrix of the true ossified tooth, and on this account can be isolated exactly like the dentinal tubes." (2) In the smaller spaces, the spherical character of the tissue composing the walls is not so apparent; the cavity has a more jagged outline, and resembles in appearance a lacuna of bone, the more so as it is traversed by the dentinal canals. In the fang, these smaller interglobular spaces and globules constitute what has been termed the 'granular layer' by Tomes. (3) True lacunæ have been rarely seen by Kölliker in dentine, and never at a distance from the cement boundary. Interglobular spaces and globules sometimes occur in the interior of the dentine of the fangs; and on the surface of the pulp cavity, the projecting spheres of dentine may produce irregularities or stalactite-like formations, visible to the naked eye. In many teeth, the interglobular spaces are absent, or but little marked: they are always most conspicuous in those specimens in which the enamel exhibits irregular development, and there appears to be a relation between the contour markings in the dentine and the grooves and irregularities in

---

(1) Op. cit., p. 255.
(2) Kölliker, op. cit., p. 294.
(3) Tomes, Lectures on Dental Physiology and Surgery, p. 48.

substances, viz. calcarious earth and an animal substance, which

---

the enamel. (1) In connection with this condition, also, Czermak has pointed out the existence of opaque white lines, forming rings round the fang: they are, of course, abnormal, and vary in breadth from the 1-50th to the 1-100th of an inch. The existence of the contour lines has been explained by referring them to the darker appearance produced by a series of secondary curves in successive dentinal tubes, or by a widening of those tubes; but it appears more probable that they are due, for the most part, to the same cause which has produced the interspaces and the globular condition of the dentine—viz., to the mode in which the animal material of dentine is calcified, and to occasional arrests in the process. Mr Salter observes that the contour markings, as well as the fracture lines, which readily occur in the intermediate normal dentine, and are parallel to them, exactly correspond to the pulp surface in the progressive formation of the dentine—are identical, in fact, with the juncture line of the pulp and internal dentine surface at any particular time of growth. The existence of interglobular spaces and persistent globules may be, with high probability, attributed to arrest at various times in the process of calcification. The observations of Czermak led him to the fact that the organic material of dentine is, during the process of calcification, impregnated with earthy salts in globular forms; and that, by a deeper process of calcific impregnation, the whole tissue is imbued with the hardening element, and the globules are fused. It is easy, therefore, by supposing a temporary interference with this process from some constitutional cause, to account for the appearances we have been describing. (2) Some light is, perhaps, thrown on the process of calcification by some observations of Mr Rainie, quoted in Mr Tomes's 'System of Dental Surgery.' "Mr Rainie finds, that if carbonate of lime is formed in a thick solution of mucilage or albumen by the decomposition of carbonate of soda or potash, the newly-formed salt takes a globular instead of a crystalline form. The globules produced are composed, however, not only of carbonate of lime, but also of a certain portion of mucilage or albumen in which the combination has taken place." Phosphate of lime, if produced under similar circumstances, supposing a small proportion of carbonate of lime be present, will also assume the globular form. The globules are laminated, and increase by the addition of new layers on the surface; and if two or more globules are in contact, they become fused into one laminated mass by the union of the laminae which are in contact. "The globules themselves have been produced by the coalescence of smaller masses, which again are made up of still smaller spherules of similar material, the individuality

---

(1) Salter, op. cit., p. 253.
(2) *Vide* Salter, op. cit.

we might suppose to be organized and vascular. The earth is in very considerable quantity; it remains of the same shape

of the constituent bodies being ultimately lost in the uniform fusion of the whole into one compact mass." "In the discovery of the substitution of the globular for the crystalline form of these two salts of lime, Mr Rainie considers he has found an explanation of the process of calcification, not only of bone and teeth, but also of the formation of shells." (1) In many teeth, the original dentine globules are indicated by faintly-traced arched lines; but the more perfectly calcified the tooth, the more completely will fusion have taken place.

The dentinal canals are generally held to be permeated during life by a nutrient fluid. Mr Tomes, however, has announced the discovery that each dentinal tube is permanently tenanted by a soft fibril, which passes from the pulp into the tube, and follows its ramifications. The dentinal fibril he describes as consisting of an almost structureless tissue, transparent, and of a comparatively low refractive power. He has not hitherto been able to determine whether it is tubular or solid. "In some cases," he writes, "there is an appearance of tubularity; but, being cylindrical, this may be a mere optical effect. When accidentally stretched between two fragments of dentine, the diameter of the fibril becomes much diminished; and when broken across, a minute globule of transparent but dense fluid may sometimes be seen at the broken end, gathered into a more or less spherical form." These appearances may lead to the surmise that the fibril, like the white fibrillæ of nerves, consists of a sheath containing a semi-fluid matter; but such a conclusion is not to be accepted without farther evidence. Mr Tomes is unable to state the manner in which the fibrils he describes are connected with the pulp. He has traced them for a short distance into its substance, but at present cannot decide whether they terminate in cells, or are in any way connected with the nerves of the pulp. Although he does not take upon himself to affirm that these fibrils are of the nature of nerves, he yet regards them as the means by which sensibility is communicated to the dentine. (2) These observations are highly interesting and important, but at present they require confirmation.

Nasmyth took a different view of the structure of dentine. From his observations, he was led to deny the existence of the dentinal tubuli. He believed that "the so-called tube was in reality a solid fibre, composed of a series of little masses succeeding each other in a linear direction, like so many beads collected on a string." The matrix—or, as he terms it, inter-fibrous substance—he describes as being originally cellular in composition; and he maintains that the baccated fibres are, in fact, rows of persistent nuclei belonging to the cells of which the inter-fibrous

---

(1) Tomes's System of Dental Surgery, pp. 298, 299.
(2) Tomes, op. cit., pp. 282-288.

after calcination, so that it is in some measure kept together by cohesion; and it is capable of being extracted by steeping in the muriatic, and some other acids. The animal substance, when deprived of the earthy part, by steeping in an acid, is more compact than the same substance in other bones, but still is soft and flexible. (r)

---

substance is composed. (1) His observations have not, however, been confirmed by those of any other writer on the dental tissues.

Dentine exhibiting Haversian canals (the vaso-dentine of Owen) is very rarely seen in Man, although it occurs in many animals. The dentine which is produced in cases of obliteration of the pulp cavity may, however, exhibit a few Haversian canals, and rounded cavities resembling *lacunæ:* this has been described by Prof. Owen under the name of *osteo-dentine*. (2)]

(r) [The organic basis of dentine (tooth cartilage) is identical chemically with that of bone: it is, by boiling, readily converted into gelatine. When separated by treating dentine with hydrochloric acid, it retains not only the form, but the internal structure of the tissue: the tubes, however, are not so easily visible. Kölliker states that if tooth cartilage "be macerated in acids or alkalies until it is quite soft, the matrix is found in the act of disintegration; but the dentinal tubes, with their walls, are still preserved, and may be isolated in large quantities." The tubes may also be isolated after long-continued boiling. By prolonged maceration in acids or alkalies, the whole of the organic basis is dissolved.

The following is the chemical composition of dry dentine according to Von Bibra: (3)

	Molar Tooth of a Man.
Phosphate of Lime and some Fluoride of Calcium	66·72
Carbonate of Lime	3·36
Phosphate of Magnesia	1·08
Salts	0·83
Cartilage	27·61
Fat	0·40
	100·00
Organic substances	28·01
Inorganic substances	71·99]

---

(1) Nasmyth, Researches, pp. 93, 94.
(2) *Vide* Kölliker, op. et loc. cit.
(3) Kölliker, op. cit., p. 292.

That part of a Tooth which is bony, is nearly of the same form as a complete Tooth; and thence, when the Enamel is removed, it has the same sort of edge, point, or points, as when the Enamel remained. We cannot by injection prove that the bony part of a Tooth is vascular: but from some circumstances it would appear that it is so; for the Fangs of Teeth are liable to swellings, (s) seemingly of the spina ventosa kind, like other bones; and they sometimes anchylose with the socket by bony and inflexible continuity, as all other contiguous bones are apt to do. (t) But there may be a deception here, for the swelling may be an original formation, and the anchylosis may be from the pulp that the Tooth is formed upon being united with the socket. The following considerations would seem to shew that the Teeth are not vascular: first, I never saw them injected in any preparation, nor could I ever succeed in any attempt to inject them, either in young or old subjects; and therefore believe that there must have been some fallacy in the cases where they have been said to be injected. Secondly, we are not able to trace any vessels going from the pulp into the substance of the new-formed Tooth; and whatever part of a Tooth is formed, it is always completely formed, which is not the case with other bones. But what is a more convincing proof, is reasoning from the analogy between them and other bones, when the animal has been fed with madder. Take a young animal, viz. a pig, and feed it with madder, for three or four weeks; then kill the animal, and upon examination you will find the following appearance: first, if this animal had some

---

(s) [Exostosis of the fang, produced by hypertrophy of the cement.]

(t) [Mr Tomes states that "although numerous instances may be found where two teeth become united by cementum developed under circumstances which constitute its formation a disease, yet in no well-authenticated instance has the cementum become continuous with the bone of the socket." (1) He believes in the existence of a law which prohibits the union of the tooth to the jaw in Man. Anchylosis of the teeth to the jaw is the normal condition in reptiles.]

(1) Tomes, op. cit., p. 445.

parts of its Teeth formed before the feeding with madder, those parts will be known by their remaining of this natural colour; but such parts of the Teeth as were formed while the animal was taking the madder, will be found to be of a red colour. This shews, that it is only those parts that were forming while the animal was taking the madder that are dyed; for what were already formed will not be found in the least tinged. This is different in all other bones; for we know that any part of a bone which is already formed, is capable of being dyed with madder, though not so fast as the part that is forming; therefore as we know that all other bones when formed are vascular, and are thence susceptible of the dye, we may readily suppose that the Teeth are not vascular, because they are not susceptible of it after being once formed. But we shall carry this still farther; if you feed a pig with madder for some time, and then leave it off for a considerable time before you kill the animal, you will find the above appearances still subsisting, with this addition, that all the parts of the Teeth which were formed after leaving off feeding with the madder will be white. Here then in some Teeth we shall have white, then red, and then white again; and so we shall have the red and the white colour alternately through the whole Tooth.

This experiment shews, that the Tooth once tinged, does not lose its colour; now as all other bones that have been once tinged lose their colour in time, when the animal leaves off feeding with madder (though very slowly), and as that dye must be taken into the constitution by the absorbents, it would seem that the Teeth are without absorbents, as well as other vessels.

This shews that the growth of the Teeth is very different from that of other bones. Bones begin at a point, and shoot out at their surface; and the part that seems already formed, is not in reality so, for it is forming every day by having new matter thrown into it, till the whole substance is complete; and even then it is constantly changing its matter.

Another circumstance in which Teeth seem different from bone, and a strong circumstance in support of their having no circulation in them, is that they never change by age, and seem never to undergo any alteration, when completely formed, but by abrasion; they do not grow softer, like the other bones, as we find in some cases, where the whole earthy matter of the bones has been taken into the constitution.

From these experiments it would appear, that the Teeth are to be considered as extraneous bodies, with respect to a circulation through their substance; but they have most certainly a living principle, by which means they make part of the body, and are capable of uniting with any part of a living body; as will be explained hereafter: and it is to be observed, that affections of the whole body have less influence upon the Teeth than any other part of the body. Thus in children affected with the rickets, the teeth grow equally well as in health, though all the other bones are much affected; and hence their Teeth being of a larger size in proportion to the other parts, their mouths are protuberant. (*n*.)

OF THE CAVITY OF THE TEETH.

Every Tooth has an internal Cavity, which extends nearly the whole length of its bony part. It opens, or begins at the point of the fang, where it is small; but in its passage becomes larger, and ends in the body of the Tooth. This end is exactly of the shape of the body of the Tooth to which it belongs. In general it may be said, that the whole of the Cavity is nearly of the shape of the Tooth itself, larger in the body of the Tooth, and

---

(*n*.) [The dentinal tubules and their ramifications, the lacunæ and canaliculi of the cement, and even the inter-fibrous spaces of the enamel, are probably all provisions by which fluid of a nutrient character may permeate the dental tissues. But there is no proper circulation through the calcified structures of the human tooth, unless in those exceptional instances in which the Haversian canals may be traced in the cement or dentine. Change of material is much slower than in bone, as is proved by the above experiments of Hunter, which have been confirmed by the observations of Flourens and others. (1) There can be no doubt, however, that the formation and development of the teeth are considerably influenced by constitutional states. Early caries having its origin in a faulty condition of the dental tissues is one of the most frequent results of a weakly, ill-nourished childhood.]

(1) Kölliker, op. cit., p. 309.

thence gradually smaller to the extremity of the fang; simple, where the Tooth has but one root; and in the same manner compounded, when the Tooth has two or more fangs.

This Cavity is not cellular, but smooth in its surface: it contains no marrow, but appears to be filled with blood-vessels, and, I suppose, nerves, united by a pulpy or cellular substance. (*v*) The vessels are branches of the superior and inferior Maxillaries; and the nerves must come from the second and third branches of the fifth pair.

By injections we can trace the blood-vessels distinctly through the whole Cavity of the Tooth; (*w*) but I could never trace the Nerves distinctly even to the beginning of the Cavity. (*x*)

(*v*) [The dentinal pulp is that portion of the fœtal tooth-papilla which remains after the development of the dentine. It is connected below with the periosteum lining the alveolus, passes up the aperture in the fang, and accurately fills the central cavity of the tooth, being everywhere in close contact with and adherent to the inner surface of the dentine. It is a soft red substance, highly vascular, with the exception of a layer on its surface, and well supplied with nerves. In structure that portion which contains vessels consists, according to Kölliker, of "an indistinctly fibrous connective tissue, with very numerous round or elongated nuclei interspersed." The same authority describes the pulp as being covered by a delicate structureless membrane, under which is seated "a layer 0·02''', 0·03''', to 0·04''' thick, consisting of several rows of cylindrical or conical nucleated cells, 0·012''' long, 0·002''' to 0·003''' broad, disposed perpendicularly upon the surface of the pulp like a columnar epithelium. Further inwards, these cells are arranged less regularly, and at length pass into the vascular tissue of the pulp without well-defined limits by the medium of shorter and more roundish cells." (1). These cells correspond with those termed by the same author "the formative cells of the dentine," and he believes that they furnish materials for any deposition of dentine on the walls of the cavity which may occur in the adult.]

(*w*) [From three to ten small arteries may be observed entering the pulp of each tooth. By ramifying in the pulp substance, they produce a loose net-work of capillaries, from which the veins arise. No lymphatics have been observed in the dentinal pulp. (2)]

(*x*) According to Kölliker, the nerve supply to each tooth consists of a principal trunk, 0·03''' to 0·04''' in size, and of six or even more fine

---

(1) Kölliker's Manual of Human Microscopical Anat., pp. 299, 300.
(2) Op. cit., p. 300.

OF THE PERIOSTEUM OF THE TEETH.

The Teeth, as we observed, are covered by an Enamel only at their bodies; but at their fangs they have a Periosteum, which, though very thin, is vascular, and appears to be common to the Tooth which it incloses, and the socket, which it lines as an investing internal membrane. (*y*) It covers the Tooth a little beyond the bony socket, and is there attached to the Gum.

twigs, 0·01‴ to 0·02‴. These ascend into the pulp cavity without forming at first considerable anastomoses, but giving off separate fibrils. On reaching the thicker part of the pulp, they form a rich plexus " with elongated meshes and collections of nerve tubes, and thus gradually break up into fine primitive fibres, 0·001‴ to 0·0016‴ in diameter." These primitive fibres form evident loops, but it is uncertain whether they represent the absolute terminations. (1) Mr Tomes states that he has been unable to trace any connection between the ultimate nerve fibres of the pulp and the dentinal fibrillæ which he describes. (2)

It would appear that at least in some instances the size of the nerve supplying the tooth-pulp is proportionate to the size and importance of the tooth. Thus Professor Owen notices the large size of the nerve supplying the laniary in the Dog and other *Carnivora*. In the Boar, he observes that the nerve supplying the developed tusk is still larger, having relation not only to the size of the tooth, but also to the continual reproduction of the matrix at its base. (3)]

(*y*) [A difference of opinion appears to exist between writers on the subject of the dental periosteum. Hunter, Bichat, and Kölliker describe but one layer of periosteum as common to the tooth and the socket. The latter author writes, "the periosteum of the alveoli is very accurately applied to the surface of the fangs." Other observers, amongst whom are Fox, Bell, Spence Bate, and Hulme, believe that the periosteum consists of two layers—one lining the alveolar cavity, the other covering the fang. Mr Bell asserts that he has frequently removed a tooth in the dead subject, and found not only the fang covered, but the socket lined with periosteum. (4) Mr Spence Bate applies the term "peridenteum" to the layer covering the fang, to distinguish it from the periosteum of the alveolus. He regards the peridenteum as a dermal tissue distinct in origin from the periosteum, and on their complete separation, he believes, depends the

---

(1) Kölliker, op. cit., p. 300.
(2) Tomes, Manual of Dental Surgery, p. 286.
(3) Odontography, Introduction, p. lxvi.
(4) The Anatomy, Physiology, and Diseases of the Teeth, by T. Bell, F.R.S., p. 43, 2nd edition.

OF THE SITUATION OF THE TEETH.

The general shape and situation of the Teeth are obvious. (z) The opposition of those of the two Jaws, and the circle which each row describes, need not be particularly explained, as they may be very well seen in the living body, and may be supposed to be already understood, from what was said of the Alveolar Processes.

We may just observe, with regard to the situation of the two rows, that when they are in the most natural state of contact, the Teeth of the Upper-Jaw project a little beyond the lower Teeth, even at the sides of the Jaws; but still more remarkably at the fore part, where in most people the upper Teeth lie before those of the Lower-Jaw : and at the lateral part of each row,

---

fact that osseous union is never observed between the tooth-fang and the alveolus. (1) Mr Hulme, in his published Lectures on Diseases of the Dental Periosteum, adduces the case of some herbivorous quadrupeds, as for instance the common Ox, in which the distinction of the layers is easily shown in support of a similar view. The latter author does not allow that the layer covering the fang is merely a reflection of that lining the socket : he believes, with Mr Bate, that there is a distinct origin for each membrane. " The alveolus," he writes, " is provided with its periosteum long before the crown of the tooth is completed; whereas in the tooth there is no periosteal layer until the formation of its fang and its investing layer of cement. At the period when the fang of the tooth is about to be formed, the sac of the toothgerm becomes adherent to the neck of the tooth; its outermost layer continues to grow in the same ratio as the fang, and becomes the formative organ of the cement. When the tooth is completed, this membrane still remains, and constitutes its periosteum. At the apex of the fully-developed fang, the periosteum of the tooth becomes intimately associated with that of the alveolus, and the contiguous surfaces of the two membranes are connected together by the passage of blood-vessels and nerves." (2) Kölliker states that the alveolar periosteum is softer than that of other bones, that it contains no elastic elements, but rich plexuses of nerves having thick nerve-tubes. (3)]

(z) [In the class Mammalia, true teeth implanted in sockets are

---

(1) On the Peridental Membrane, in its relation to the Dental Tissues,—British Journal of Dental Science, vol. i., p. 7.

(2) Dental Review, vol. iii., p. 203.

(3) Kölliker, op. cit., loc. cit.

the line, or surface of contact, is hollow from behind forwards, in the Lower-Jaw; and in the same proportion it is convex in the Upper-Jaw.

The edge of each row is single at the fore part of the Jaws; but as the Teeth grow thicker backwards, it there splits into an internal and external edge. The canine Tooth, which we shall call *Cuspidatus*, is the point from which the two edges go off; so that the first grinder, or what we shall call the first *Bicuspis*, is the first Tooth that has a double edge.

OF THE NUMBER OF TEETH.

Their number in the whole, at full maturity, is from twenty-eight to thirty-two: I once saw twenty-seven only, never more than thirty-two. Fourteen of them are placed in each Jaw, when the whole number is no more than twenty-eight; and sixteen, when there are thirty-two. If the whole be twenty-nine or thirty-one in number, the Upper-Jaw sometimes, and sometimes the Lower, has one more than the other; and when the Number is thirty, I find them sometimes divided equally between the two Jaws; and in other subjects sixteen of them are in one Jaw, and fourteen in the other. In speaking of the Number of Teeth, I am supposing that none of them have been pulled out, or otherwise lost; but that there are from eight to twelve of those large posterior Teeth, which I call Grinders, and that they are so close planted as to make a continuity in the circle: and in this case, when the number is less than thirty-two, the deficiency is in the last grinder. (*a*)

---

restricted to the maxillary, premaxillary, and inferior maxillary bones. They form a single row in each jaw, and in most Mammals, as in Man, occupy each of the above-mentioned bones. They may, however, project only from the premaxillary bones, as in the Narwhal; or only from the lower maxillary bone, as in *Ziphius*; or be implanted only in the superior and inferior maxillaries, and not in the premaxillaries, as in the true Ruminantia and most Bruta (Sloths, Armadillos, Orycteropes). (1)]

(*a*) [The normal number of the permanent teeth in Man is thirty-two.

(1) Owen on the Characters of the Class Mammalia,—Journal Proc. Lin. Soc., vol. ii., p. 6.

The Teeth differ very much in figure from one another; but those on the right side in each Jaw resemble exactly those on the left, so as to be in pairs; and the pairs belonging to the Upper-Jaw nearly resemble the corresponding Teeth of the Lower-Jaw in situation, figure, and use.

Each Tooth is divided into two parts, viz., first, the body, or that part of it which is the thickest, and stands bare beyond the Alveoli and Gums; secondly, the fang, or root, which is lodged within the Gum and Alveolar Process: and the boundary between these two parts, which is grasped by the edge of the Gum, is called the Neck of a Tooth. The bodies of the different Teeth differ very much in shape and size, and so do their roots. The difference must be considered hereafter.

The Teeth of each Jaw are commonly divided into three classes, viz. *Incisors, Canine,* and *Grinders;* but from considering some circumstances of their form, growth and use, I chuse to divide them into the four following classes, viz. *Inci-*

It occasionally happens that the last molars, or wisdom teeth, remain enclosed in their bony recesses until a late period of life. In some few instances they may be absent.

In the class Mammalia, examples of the greatest number of teeth are observed amongst those forms which generate but a single set of teeth *(Monophyodonts)*, and occur in the orders Cetacea and Bruta. Thus the priodont Armadillo has ninety-eight teeth; the Cachalot, upwards of sixty, most of them being confined to the lower jaw; the common Porpoise has between eighty and ninety; the Gangetic Dolphin, one hundred and twenty; whilst the true Dolphin presents the maximum number in Mammalia, having from one hundred to one hundred and ninety teeth. Where the teeth are excessive in number, they are of small size, equal or nearly so, and usually conical in form.

In placental Mammalia which have two sets of teeth the one succeeding the other *(Diphyodonts)*, the number of the permanent teeth never exceeds forty-four.

A few genera and species of Mammalia are strictly edentulous. Thus the true Ant-eaters *(Myrmecophaga)*, the scaly Ant-eaters *(Manis)*, and the spiny monotrematous Ant-eaters *(Echidna)* possess no true teeth, although in *Echidna* horny processes are present analogous to the lingual and palatal teeth in fishes. (1)]

---

(1) Owen, op. cit., pp. 5—9, and Art. Odontology, Enc. Brit., 8th edit., p. 438.

sores, commonly called Fore Teeth; *Cuspidati*, vulgarly called Canine; *Bicuspides*, or the two first Grinders; and *Molares*, or the three last Teeth. The number of each class, in each Jaw, for the most part, is four *Incisores*, two *Cuspidati*, four *Bicuspides*, and four, five, or six *Molares*. (*b*)

There is a regular gradation, both in growth and form through these classes, from the *Incisores* to the *Molares*, in

---

(*b*) [The permanent human dentition is indicated by the subjoined formula:

$$i\frac{2-2}{2-2} \quad c\frac{1-1}{1-1} \quad p\frac{2-2}{2-2} \quad m\frac{3-3}{3-3} = 32$$

The following is the definition and classification of teeth in diphyodont Mammalia proposed by Professor Owen. The permanent teeth in diphyodont Mammalia are referrible to four classes—viz., incisors, canines, premolars (called in Man bicuspides), and molars. The terms 'incisor,' 'canine,' and 'molar' had originally reference to the shape and office of the teeth: they are now, however, used in Comparative Anatomy to distinguish teeth according to their relative position, connections, and development; the latter circumstances constituting a truer guide to the homologies of organs than shape, size, or office. The incisors are those teeth which are implanted in the premaxillary bones and in the corresponding part of the lower jaw. Thus the term incisor comes to include the tusk of the Elephant, the pectinated lower front teeth of the Flying Lemur, the broad tuberculate front teeth of some Carnivora. The tooth in the superior maxillary bone which is situated at or near to the suture between it and the premaxillary, is the upper canine; the lower canine is that tooth which, in opposing the upper, passes in front of its crown when the mouth is closed. The premolars are those which succeed the deciduous molars. The permanent teeth, the most posterior in position, which do not occupy places previously occupied by deciduous teeth, but which are themselves a backward continuation of the first series of teeth, being developed in the same primary groove of the fœtal gum, are the true molars.

From the dentition of early forms of both Herbivorous and Carnivorous Mammalia which existed during the eocene tertiary periods, Professor Owen has been led to regard three incisors, one canine, and seven succeeding teeth on each side of both jaws as the type formula of diphyodont dentition. Of the seven teeth, four may be premolars succeeding four deciduous molars, and three molars; or there may be three premolars succeeding three deciduous molars, and four true molars. The latter type of dentition is peculiar to the Marsupials and Monotremes; the former is the typical formula of the placental diphyodont series. Amongst the few existing Mammals which have retained

which respect the *Cuspidati* are of a middle nature, between the *Incisores* and *Bicuspides*, as the last are between the *Cuspidati* and *Molares*; and thence the *Incisores* and *Molares* are the most unlike in every circumstance.*(c)

---

* It is here to be understood, that the Teeth from which we take our description, are such as are just completely formed, and therefore not in the least worn down by mastication. Our description of each class is taken from the Lower-Jaw; and the difference between them, and their corresponding classes in the Upper immediately follows that description.

---

the typical number of teeth, are the Hog and the Mole. The permanent teeth in the genus *Sus* are indicated by the following formula:

$$i\frac{3-3}{3-3} \quad c\frac{1-1}{1-1} \quad p\frac{4-4}{4-4} \quad m\frac{3-3}{3-3} = 44$$

" When the premolars and the molars are below this typical number, the absent teeth are missing from the fore part of the premolar series and the back part of the molar series. The most constant teeth are the fourth premolar and the first true molar; and these being known by their order and mode of development, the homologies of the remaining molars and premolars are determined by *counting the molars from before backwards*,—*e.g.*, 'one,' 'two,' 'three,'—and the premolars *from behind forwards*—'four,' 'three,' 'two,' 'one.' The incisors are counted from the median line, commonly the foremost part, of both upper and lower jaws, outwards and backwards. The first incisor of the right side is the homotype, transversely, of the contiguous incisor of the left side in the same jaw, and vertically, of its opposing tooth in the opposite jaw; and so with regard to the canines, premolars, and molars; just as the right arm is the homotype of the left arm in its own segment, and also of the right leg in a succeeding segment. It suffices, therefore, to reckon and name the teeth of one side of either jaw in a species with the typical number and kinds of teeth: *e.g.*, the first, second, and third incisors; the first, second, third, and fourth premolars; the first, second, and third molars; and of one side of both jaws in any case." (1) The following formula therefore represents the homologies of the human teeth in relation to the typical formula: *i.* 1, *i.* 2; *c*; *p.* 3, *p.* 4; *m.* 1, *m.* 2, *m.* 3;—the third incisor and the first and second premolars being suppressed in Man, as they are also in the Catarhine Quadrumana. (2)]

(c) [The large size of the molars as compared with the incisors, canines, and premolars, and the regular gradation in size of the human teeth from the incisors to the molars, are amongst the dental characteristics which distinguish Man from the higher Apes.]

---

(1) Owen, op. cit., p. 9.
(2) Owen, Art. Teeth, Todd's Cyclopædia, p. 904.

OF THE INCISORES.

The *Incisores* are situated in the anterior part of the Jaw ; (*d*) the others more backwards on each side, in the order in which we have named them. The bodies of the *Incisores* are broad, having two flat surfaces, one anterior, the other posterior. These surfaces meet in a sharp cutting edge. The anterior surface is convex in every direction, and placed almost perpendicu-

---

(*d*) It has been already stated that in scientific nomenclature the term incisor is restricted to those teeth which are implanted in the premaxillary bones, and in the corresponding part of the lower jaw. In the Mammalian series, (1) if we commence with the order *Bruta*, we find many instances of the total absence of teeth thus implanted. This is the case in the Phyllophagous *Bruta*—the Sloths or Tardigrades; and amongst the Insectivorous *Bruta*, in the Cape Ant-eater (*Orycteropus Capensis*), and in the Great Armadillo (*Priodon gigas*). In the Armadillos of the sub-genus *Euphractus*, however, a single tooth which resembles the succeeding molars is implanted in the premaxillary bone ; and in the lower jaw the two anterior teeth being in advance of the premaxillary tooth, are also to be considered incisors. A portion of the lower jaw of the great extinct Armadillo (*Glyptodon*) shows that the teeth extend close to the symphysis, a corresponding implantation in the premaxillaries above being thereby indicated.

The premaxillary bones are generally toothless in the true *Cetacea*: the Dolphin, however, is an exception, one pair of its numerous teeth being premaxillary. The Narwhal (*Monodon monoceros*) may be also considered another exceptional instance. In the female Narwhal, a tusk is concealed in each premaxillary bone at its junction with the maxillary, but remains throughout life an abortive germ; in the male, the tooth on the left side continues to grow, and is developed into the so-called "horn," acquiring a length of nine or ten feet.

In common with many *Bruta*, the Carnivorous *Cetacea* exhibit an inferiority in their dental system not only in the acquisition of but a single set, but in the uniformity of shape which characterises the dental organs ; but on turning to the Aquatic Pachyderms, designated by Cuvier *Herbivorous Cetacea*, we find a higher type of dentition attained. The Dugong (*Halicore*) has incisors distinguished both by shape and position from the molars ; in the upper jaw are two deciduous incisive tusks

---

(1) In compiling the information contained in the above and following notes on the Comparative Anatomy of the Teeth in Mammalia, the Editor has been chiefly indebted to the writings of Prof. Owen, F.R.S., especially the article 'Odontology,' in the last edition of the ' Encyclopædia Britannica.'

larly; and the posterior is concave and sloping, so that the cutting edge is almost directly over the anterior surface.

These surfaces are broadest and the Tooth is thinnest at the cutting edge, or end of the Tooth, and thence they become gradually narrower and the Tooth thicker towards the neck,

which are displaced vertically and succeeded by permanent tusks. It is, however, in the male sex and in the upper jaw only that these teeth project from the gum; in the female Dugong both upper and lower incisors remain concealed throughout life. The superior incisors are two in number in both sexes; in the male the extremity only of the tusk projects from the jaw, at least seven-eighths of its extent being lodged in the socket. The male tusk is subtrihedral, moderately and equally curved, and its extremity is bevelled off to a cutting edge like the scalpriform incisors of the Rodents. In the young American Manatee (*Manatus Americanus*) each premaxillary bone supports a deciduous tusk, which, however, is not replaced. In the gum which covers the deflected portion of the ramus of the lower jaw in the new-born Manatee, six depressions for rudimental teeth are to be observed; in one of these an incisor tooth was observed by Stannius.

Incisors are present in all the species of the order *Marsupialia*, but they vary in number, in some instances exceeding that of the Mammalian type. In the carnivorous Dog-headed Thylacine and Ursine Dasyure, there are eight incisors in the upper jaw, and six in the lower; in the Dasyures these teeth are simple in structure, and are arranged in a regular semicircle. The existing Australian genus *Myrmecobius*, and the extinct Marsupial genera *Phascolotherium* and *Amphitherium*, found in the oolitic slate at Stonesfield in Oxfordshire, afford examples in which the incisors are separated from each other and from the canines by vacant spaces. Ten incisor teeth in the upper jaw and eight in the lower are found in the Opossums (*Didelphys*); the two middle upper incisors are longer than the others, from which they are separated by a short interspace. In the Tapoa (*Phalangista fuliginosa*), the Koala (*Phascolarctos cinereus*), the Kangaroo Rats (*Hypsiprymnus*), and the Kangaroos (*Macropus*), there are six incisors in the upper jaw, whilst the lower are reduced to two. The two anterior upper incisors are more than twice as large as the lateral in the Koala; the same teeth also attain large dimensions as compared with the others of the series in the Kangaroo Rats: in the latter genus their pulps are persistent. The feeble development of the lateral superior incisors in *Phascolarctos* and *Hypsiprymnus* is succeeded by their total suppression in the Marsupial Rodent, the Wombat. In the Wombat (*Phascolomys*), the incisors are two in the upper and two in the lower jaw; they are genuine *dentes scalprarii* with persistent pulps, although inferior, especially in the lower jaw, in length and curvature to those of the true Rodents

where the surfaces are continued to the narrowest side, or edge of the fang. The body of an *Incisor*, in a side-view, grows gradually thicker, or broader, from the edge or end of the

---

A transition between the dentition of the Kangaroos and the Wombat is presented by a gigantic extinct Australian herbivorous Marsupial, the Diprotodon. In this species the general dentition approached that of the Kangaroo; but the median upper incisors were large curved scalpriform tusks which worked against a pair of procumbent tusks below.

In the different species of the order *Insectivora*, the incisors differ in number and size: in some species the anterior ones approximate more or less to the scalpriform teeth of the Rodents. In the Cape Mole (*Chrysochloris Capensis*), the Shrew Moles of America (*Scalops*), in the *Solenodon paradoxus* of Hayti, there are three incisors in each premaxillary, and the median one is of large size. In *Scalops* this tooth is scalpriform, in *Chrysochloris* and *Solenodon* laniariform. In the lower jaw the anterior incisor in these genera is of small size and procumbent, whilst the second incisor is large and laniariform: a third lower incisor of small size is present in *Chrysochloris* and *Solenodon*, but is absent in *Scalops*. The incisor teeth in the European Mole (*Talpa*) are six in number in each jaw, small, and simple in conformation; the fourth tooth on each side below, although resembling the incisors in shape, is to be considered a canine—its crown passing in front of the upper canine when the mouth is shut. The typical Shrews (*Soricidæ*) manifest their analogy to the Rodents by the great preponderance in size of the first two incisors in both jaws, and the sub-genera of Shrews are partly founded on variations in the shape of these teeth. In the European Hedgehog (*Erinaceus Europæus*) these teeth are six in each jaw, the anterior pair both above and below being larger and longer than the rest and very deeply implanted. The number is reduced to; four in each mandible in the tropical Hedgehogs *Echinops* and *Ericulus;* in the Tenrecs or Tailless Hedgehogs of Madagascar (*Centetes*) there are four small incisors above, whilst the typical number six is retained below.

The incisors in *Cheiroptera* may be entirely wanting or present in the numbers 1.1. to 2.2. in the upper jaw, and 1.1. to 3.3. in the lower; they are always very small, and in the upper jaw unequal and separated by a median interval. In the Suctorial or Vampire Bats (*Desmodus*) the anterior teeth are modified in accordance with their habits. The upper incisors are two in number, one in each premaxillary bone, closely approximated, with a very large compressed curved and sharp-pointed crown, implanted by a strong fang which extends beyond the premaxillary into the maxillary bone. They are succeeded by similarly-formed canines. In the lower jaw, the incisors, two in number on each side, are much smaller than in the upper, and have bilobed crowns. In

Tooth to its neck; and these coincide with the flat, or broad side of the fang; so that when we look on the fore part, or on the back part of an *Incisor*, we observe it grows constantly

---

the Colugos (*Galeopithecus*), which in their general organisation approach the Lemurs, although placed by Cuvier at the end of the Bats, we find two incisors in the upper jaw on each side, of which the two anterior are separated by a wide interval; in the Philippine Colugos they are small with bilobed crowns. In *Galeopithecus Temminckii* the crown is an expanded plate with three or four tubercles; the second upper incisor in both species is implanted by two fangs. The lower incisors are three in number on each side. The crowns of the first two present the form of a comb,—a configuration which, unique in Mammalia, depends upon the prolongation of notches deeper and more numerous, but yet analogous to the indentations on the cutting edge of the newly-formed human incisor.

In the extensive order of Rodents, we find the incisors represented by a large curved ever-growing pair of teeth in each jaw, which are separated by a wide interval from the short molar series. To this there is but one exception—the family of the *Leporidæ*, Hares, Rabbits, and Picas, in which a minute second incisor occurs on each side behind the large upper scalpriform tooth. The upper incisors describe a larger segment of a smaller circle; the lower, a smaller segment of a larger circle: the lower are the larger teeth, and their sockets extend to the back part of the jaw on the inner side of the molars. These teeth are of unlimited growth; and if by accident the opposing teeth are lost, they continue to grow until they describe a complete circle, perforating the skin, and causing absorption of the bone with which they come in contact. Throughout life the base of the tooth continues widely open, and contains a long conical persistent pulp, which is surrounded by a progressively ossifying capsule. Near the crown an enamel-pulp is attached to the inner side of that part of the capsule which covers the convex surface of the curved incisor. The tooth consists of a body of dentine with a plate of enamel on its anterior surface, and a general investment of cement, which is, however, very thin upon the enamel. The mutual action of the lower and upper incisors produces an obliquely worn surface on the extremity of the crown, which slopes backwards from the anterior edge of hard enamel: the resemblance so produced to the cutting edge of a chisel has given the name of "*dentes scalprarii*" to these teeth. The varieties the scalpriform incisors present in the different species of Rodents are limited to their size, and the colour and sculpturing of their anterior surface. In the Coypu, Beaver, Agouti, and some other species, the enamelled surface is of an orange or brownish-red colour. In some genera the same surface presents a longitudinal groove.

narrower from its cutting edge to the extremity of its fang. But in a side-view it is thickest or broadest at its neck, and thence becomes gradually more narrow, both to its cutting edge, and to the point of its fang.

On turning to the order *Quadrumana*, we find a genus (*Cheiromys*) in which, as in the Marsupial Wombat, the Suctorial Bats, and the Insectivorous Shrews, the dentition is modified in analogy with the Rodent Type. In the Aye-aye (*Cheiromys Madagascariensis*) canines are wanting, and a wide interval separates two large curved scalpriform incisors from the molar series. These teeth differ from the incisors of the Rodents in having an entire investment of enamel, which is, however, thickest on the anterior surface. In the slow Lemurs (*Stenops*), and in the true Lemurs, or Makis, there are four incisor teeth in each jaw. In the *Stenops tardigradus* the first upper incisor is larger than the second: in the true Lemurs the upper incisors are small and vertical, with expanded crowns; an interval separating the two on the right side from those on the left. The inferior incisors with the canine are long, narrow, compressed, and procumbent. In the Platyrhine Quadrumanes of America, as in the Quadrumana of the Old World, the incisors are $\frac{2}{2}$: $\frac{2}{2}$: ; the Sakis manifest the lemurine character of long narrow inferior incisors. In *Cebus* the inferior incisive teeth have broad, thick, wedge-shaped incisive crowns; and this character is generally retained throughout the Quadrumanous series. The *Simiæ* of the Old World present the nearest approach to human conformation offered by the animal kingdom. The large black tailless Apes of Africa (*Troglodytes*), and the red Orangs of the Indian Archipelago (*Pithecus*), afford the closest approximation. It will be interesting here to compare this part of the dental series in the two best-known African species—the Gorilla and Chimpanzee, and in the two species of Orang, with the human permanent incisors.

The characteristics of the human incisors are their smallness of size in proportion to the other teeth and to the entire skull; their near equality of size with each other; their thin wedge-like form, and their vertical, or nearly vertical, implantation. In the upper jaw the two middle incisors slightly exceed the lateral in breadth; the posterior surfaces of both are smooth and slightly concave. The middle incisor presents a slightly-developed basal ridge on its posterior surface; but such elevation is entirely absent on the posterior surface of the lateral one. The external angle of the crown of the lateral incisor is not rounded off, and is in contact with the canine. In the lower jaw the incisors are somewhat smaller than in the upper, and the breadth of the central is slightly inferior to that of the lateral; the outer angle of the crown of the latter is entire. In the Gorilla (see plate I) and Chimpanzee the incisor teeth are not only absolutely much larger (thicker and broader) than in man, but they are of far greater size in proportion to the molar series

The Enamel is continued farther down, and is thicker on the anterior and back part of the *Incisores* than on their sides, and is even a little thicker on the fore part than upon the back part of the Tooth. If we view them laterally, either when intire, or

---

and to the entire skull. In the Gorilla, the transverse measurement of the four incisor teeth is not greater than in the Chimpanzee; they are therefore proportionately smaller; and in this respect, although still very far removed, the Great Ape makes a nearer approach to human dimensions. A further deviation from human type is seen in the greater inequality of the incisors of the upper jaw, the central incisors being considerably broader than the lateral. This inequality is greatest in the Gorilla, but in both species it is a marked point of differentiation. Each incisor has a well-developed prominent basal ridge on its posterior surface; and the outer angle of the crown of the lateral, instead of being entire, as in Man, is rounded off. In the lower jaw these teeth are equally characterised by their great size. As in Man, the lower lateral are broader than the central ones; but the former have the external angles of the crowns rounded off. In both jaws the direction of the incisive teeth departs from a vertical or nearly vertical position. In the incisors of the Orangs we notice a still greater divergence from human type in the superior breadth of the entire series, in the greater size of the upper central teeth, and the greater amount of inequality between them and the lateral. The central incisor of the upper jaw in *Pith. Satyrus* and *Pith. Morio* is of unusual size and strength, being double the width of the outer one, whilst its thickness is nearly equal to its transverse measurement. From their constant use in overcoming the resistance of hard and tough vegetable substances, they become worn down; and in the old Orang they present a broad abraded surface, which extends obliquely backwards from the cutting edge to the posterior margin of the base of the crown. The lateral incisors have their external angles obliquely truncated; in the lower jaw they exceed in transverse measurement the dimensions of the central. In the obliquity of their implantation in both jaws they exhibit the ordinary Quadrumanous character. In the Gorilla and Great Orang, as in the Quadrumana generally, a well-marked interval or diastema separates the upper incisors from the canine.

The incisors are six in number in each jaw in the typical Carnivora. In the Lion they form a transverse row; the outermost above is the largest, and resembles a small canine; the intermediate ones have broad and thick crowns indented by a transverse cleft. These teeth are employed in gnawing the gristly ends of bones, and scraping off the attachments of periosteum and muscles. The crowns of the incisors in the Hyænas form a similar transverse line; the crown of the external one above is laniariform; the crowns of the intermediate teeth are

when cut down through the middle, but especially in the latter case, it would seem as if the fang was driven like a wedge into, and had split the body or Enamel of the Tooth. They stand

---

divided by a transverse cleft into a strong anterior conical lobe and a posterior ridge, which is notched vertically, giving to the tooth a three-lobed configuration. The inferior incisors increase in size from the first to the third; the second and third have the crown indented externally, but they do not present a posterior notched ridge; the conical points of these teeth fit into the depressions separating the lobes of the incisors above. The incisive teeth in the Dog form a segment of a circle in both jaws; they increase in size from the first to the third; the edge is divided by two notches into a large middle and two smaller lateral lobes. In the most aquatic and piscivorous of the *Mustelidæ*—viz., the Sea-Otter (*Enhydra*)—the number of incisors is reduced by the absence of the two central in the lower jaw; their number is still farther diminished in some of the *Phocidæ:* in the great proboscidian and hooded Seals (*Cystophora*) the incisor formula is $\frac{4 4}{2 2}$; they are, however, of large size, and laniariform, the two outer ones above being largest. In the young Walrus there are three teeth in each premaxillary above, and two on each side below; but they soon disappear, with the exception of the outer and upper one on each side, which remains on the inner side of the enormous canine tusk.

The normal number of incisive teeth, $\frac{3 3}{3 3}$, which we find in the extinct herbivorous Dichodon of the eocene tertiary deposits, is replaced in the existing typical Ruminants by the total suppression of incisors in the upper jaw, whilst the number six is retained in the lower. In the hollow-horned Ruminants *Antilopidæ*, *Ovidæ*, and *Bovidæ*, a callous pad supplies the place of the upper incisors, albeit their rudiments have been observed by Professor Goodsir and others in the embryo Sheep and Cow. The incisors in the lower jaw have low, broad crowns. Upper incisors are also absent in the solid-horned *Cervidæ:* in the Camels one laniariform incisor is present in each premaxillary bone. Professor Owen has described six deciduous upper incisors in the new-born Dromedary, of larger size than any rudiments of these teeth which exist in the hollow-horned Ruminants.

The *Suidæ*, or Hog tribe, retain the typical dental formula which is exemplified in the *Chœropotamus*, *Anthracotherium*, and other extinct non-ruminant Artiodactyles. In the Hog, the upper incisors decrease in size from the first to the third; the central upper incisors are inclined to each other, and touch by their prolonged inner surfaces; their crowns are short, strong, and obtusely pointed: the crown of the second is as broad as that of the first, but shorter and thinner; it has a trenchant and dentated edge, which soon, however, becomes worn down. The third is a small tooth separated by a short interval from the second. In the

almost perpendicularly, their bodies being turned a little forwards. Their fangs are much shorter than those of the

---

lower jaw, the incisors are long, straight, and sub-compressed: the third is the smallest, the second rather larger than the first. An excessive development of the canine and incisor teeth marks the genus *Hippopotamus*. Of the incisors there are four in each jaw; the lower attain the most remarkable development. The two median lower incisive tusks are cylindrical, of great size and length, and are worn obliquely on the outer and upper side of their extremity; the deeply-implanted basal portion is grooved longitudinally: the two outer incisors are likewise cylindrical and straight, but of smaller size; they exhibit an abraded surface towards the inner side of the apex. A large persistent matrix occupies the excavated base of these teeth, and provides for constant growth and reparation. An ancient form of Hippopotamus which formerly inhabited India retained the typical number of incisors $\frac{4}{4}:\frac{4}{4}:$.

Amongst Perisso-dactyle *Ungulata* that number is retained in the genera *Equus* and *Tapirus*. In the Horse, the crowns of the incisors form the arc of a circle at the extremity of each jaw. They are distinguishable from the incisors of all other animals by the vertical fold of enamel which dips down into the substance of the crown from its broad, flat upper surface, like the inverted finger of a glove. In the moderately-worn tooth, the fold of enamel remains surrounded by a cavity, which is partly filled by cement and partly by the *débris* of the food. This constitutes the "mark" of horsedealers. The mark disappears in the aged animal when the tooth becomes worn still lower than the fold of enamel. In the different species of Rhinoceros, the incisor teeth present considerable variation. They may be absent, and when present they differ much in form and proportions, and it has been remarked that their development has a close relation to the development of the nasal weapon. The two-horned Rhinoceroses of Africa, which have one or both of the horns largely developed, and a great extinct species (*Rh. tichorinus*) in which the horns attained a prodigious size, are instances of the total absence of these teeth in the adult condition. The Sumatran two-horned Rhinoceros, in which the horns are but moderately developed, exhibits incisors in both jaws; they also exist in both the one-horned species; but these teeth attained their largest development in an extinct hornless Rhinoceros (*Rh. incisivus*.) The Tapir presents six incisors in each jaw. The upper median has a broad trenchant crown separated by a transverse channel, into which the wedge-shaped crown of the lower incisor fits, from a basal ridge. The outermost pair of the upper incisors are large and canine-like, the lower are unusually small.

In the extinct genus *Dinotherium* the incisors are represented by two large tusks implanted in the deflected extremity of the lower jaw. The crowns of these tusks gradually decrease to a point; they are bent

Cuspidati, but pretty much of the same length with all the other Teeth of this Jaw. (c)

In the Upper-Jaw they are broader and thicker, especially the two first: their length is nearly the same with those of the Lower-Jaw. They stand a little obliquely, with their bodies turned much more forwards (the first especially) and they generally fall over those of the Under-Jaw.

The two first *Incisores* cover the two first, and half of the second of the Lower-Jaw, so that the second *Incisor* in the

---

downwards and backwards; the base, like that of the tusk of the Elephant and Mastodon, is widely excavated for a persistent pulp. No corresponding teeth have been discovered in the upper jaw. In the existing Proboscideans, two deciduous tusks in the upper jaw, replaced by a pair of permanent tusks of large size, are the sole representatives of incisors; but in the extinct Mastodons, two lower incisive tusks were also present, although it is uncertain whether they were preceded by deciduous teeth. The tusk of the Elephant, meeting no opponent to its growth, increases to an enormous length, following the curve originally impressed upon it by the form of its socket. Its growth may be compared to the abnormal increase of the scalpriform incisor of the Rodent when accident has deprived it of an opposing tooth. In the female of the Indian species, the tusk is shorter and straighter, and less deeply implanted, than in the male. According to Cuvier, large tusks are present in both sexes in the African Elephant; at least, this is the case in certain localities. Professor Owen is of opinion that the smaller fossil tusks of the Mammoth (*Elephas primigenius*) which have been discovered belonged to the females of that extinct species. (1)]

(e) [A degree of departure from the perpendicular direction in the incisor teeth, especially in the upper, is a usual concomitant of the prognathic conformation; and although most common in the Melanian races, is occasionally observed in every variety of Man. Amongst the African nations, the prognathic development of the maxillæ, and the consequent obliquity of implantation of the incisor teeth, appear to be most remarkably expressed in the tribes of the West Coast. The Quaiquœ race, including the Bushmen and Hottentots, exhibit the prognathic character in a less degree than do the true Negroes. (2)]

---

(1) Prof. Owen, Art. Odontology, Encyclopdæia Britannica, 8th edit., vol. xvi., pp. 440-477.

(2) 'On the Teeth in the Varieties of Man,' DENTAL REVIEW, 1860.

Upper-Jaw covers more than half of the second, and more than the half of the *Cuspidatus* of the Under-Jaw.

The edges of the *Incisores* by use and friction, in some people, become blunt and thicker; and in others they sharpen one another, and become thinner. (*f*)

OF THE CUSPIDATUS.

The *Cuspidatus* is the next after the *Incisores* in each Jaw; so that there are four of them in all. (*g*) They are in general

---

(*f*) In barbarous and semi-civilised races, the incisors and canines become excessively worn down by the rough work to which they are put. The Esquimaux, the Fuegians, and the Oceanic nations, including both Malayo-Polynesian and Negrito varieties, exemplify remarkably the effects of attrition on these teeth. The same condition is frequently seen in Anglo-Saxon and Early British crania. The appearance of wearing down of these teeth is also very remarkable in many Ancient Egyptian and Guanche mummies. Blumenbach describes the incisors in Egyptian mummies as thick and round — not, as usual, flattened into edges, but resembling truncated cones; and the cuspidati are not pointed, but broad and flat on the masticating surface, like the neighbouring bicuspides. Mr Lawrence, who investigated the subject, came to the conclusion that this condition did not depend on any natural variety; and an examination of the teeth in two mummies containing the remains of children whose death had occurred between the completion and loosening of the first set of teeth, instituted by Dr Prichard and Mr Estlin, proved that the original conformation of the teeth in the Ancient Egyptian was in exact accordance with the normal type. There is, it may be added, no difference in the characters of the incisive teeth in the varieties of Man. In the Negro races, they are usually large, broad, and thick; but they are not of greater absolute dimensions than in numerous individuals among the leucous and xanthous varieties. (1)]

(*g*) [The term canine has been already limited to that tooth in the superior maxillary bone, which is situated at or near to the suture between it and the premaxillary, and in the lower jaw to that tooth which, in opposing the upper canine, passes in front of it when the mouth is closed. Neither the insectivorous or phyllophagous *Bruta* possess true canines. The two-toed sloth (*Cholœpus didactylus*), however, differs from the other genera of the order, in having the first

---

(1) Lawrence, History of Man, 9th edit., p. 260; Prichard, Physical History of Mankind, vol. ii., p. 249; DENTAL REV., loc. cit.

thicker than the *Incisores*, and considerably the longest of all the Teeth.

---

tooth in both jaws of larger size than the succeeding ones; and, from the peculiarity of their shape, these larger teeth have received the name of canine. An interval, which is wider in the upper jaw, separates the so-called canine from the succeeding tooth; but, in consequence of the greater interspace above, the upper tooth plays upon the anterior surface of the lower, thus the relative position which characterises the true canines in Carnivora and Quadrumana is reversed. These teeth are three-sided, with the margins of the abraded surfaces leading to the point, trenchant.

The presence and devolopment of canines in the Marsupial order, evidently present a close connection with the food and habits of its different genera. Thus, in the predaceous Thylacinus and Dasyurus, these teeth resemble closely in shape and proportions, the same weapons in the typical Carnivores. In the Dog-headed Opossum (*Thylacinus*) the canines are long, strong, curved, and pointed: the apices of the lower canines are received in depressions in the pre-maxillary palatal plate when the mouth is closed, and do not, as in the placental carnivores, project beyond the margin of the upper jaw. In *Dasyurus*, the canines are of equal, or even greater development than in the Thylacine. In an extinct animal of this genus, they resembled in form and proportions the same teeth in the Leopard. The canines present a diminished relative size in the Phascogale and other smaller species of the carnivorous group. The Opossums (*Didelphys*), present a dentition adapted to the mixed nature of their diet. Canines of considerable developuent are present in both jaws; but the molars indicate a departure from the conformation characteristic of the true flesh-feeders. The smaller species of Opossums are insectivorous; whilst the larger prey on small quadrupeds and birds, as well as on reptiles and insects, and even fruits. One species prowls along the sea-shore in search of crabs and other crustacea; another, Otter-like, subsists almosts exclusively on fresh-water fish. Canines are feebly represented in the Phalangers and Petaurists, which feed principally on fruits, buds, and leaves, together with insects, eggs, &c. In the herbivorous Potoroos and in the Koala, they are absent in the lower jaw; whilst they are wanting in both jaws in the Kangaroos and Wombat.

The large tailless Hedgehogs, or Tenrecs of Madagascar, have the most formidably-developed canines to be found in the whole Insectivorous order. In the Tenrec, these teeth present all the characters of the canines of Carnivora: they are large and long, pointed, trenchant, recurved, and single-fanged. A wide space separates them in both jaws from the premolars. In the European Hedgehog, no canine teeth are present. The first tooth which succeeds the incisors in the *Gymnura*, in both jaws,

The shape of the body of the Cuspidatus may be very well

resembles in form and size a canine; but in the upper jaw it is implanted by two roots. In the Tupaias (*Glisorex*), small canines are present in both jaws. None of the small premolars which succeed the incisors in the Shrews fulfil the office, or have the configuration of a canine. The common Mole (*Talpa Europœa*), has an upper canine of large size, with a crown of characteristic shape; but it is implanted by two fangs. The tooth in the lower jaw, which by right of position is to be considered the canine, is small, simple, and resembles the incisors; whilst the first premolar has a laniariform crown, and like the canine above, has a double implantation. This latter tooth is considered the canine by Bell and Dr Blainville. In *Talpa mooguru* the inferior canine is absent. In the *Condylure*, or Rayed-Mole, it is present, and has the form and proportions of a canine. Canines are absent in the lower jaw in the genus *Scalops*. Neither by virtue o shape or office are the teeth which succeed the incisors in the Chrysochlore to be considered canines.

Canines of normal form are present in both jaws in the Bats. In the Vampire (*Desmodus Vampirus*), the canines have large, lancet-shaped crowns similar to those of the incisors.

The absence of canines is one of the characteristics of Rodent dentition.

Order *Quadrumana*. In the Aye-aye (*Cheiromys*), whose dentition makes a near approach to that of the Rodents, canines are absent. They are present in the Slow Lemurs (*Stenops*), and in the True Lemurs, or Makis. In the True Lemurs, the inferior canines are compressed, and procumbent like the incisors. The upper canine tooth is long, curved, compressed, sharp-edged, and pointed. Canines are present in both jaws in the Platyrhine, as in the Catarhine *Quadrumana*. They are formidable weapons in the Capuchin Monkeys (*Cebus*). In the Monkeys and Apes of the Old World they are always longer than the adjoining teeth; they are pointed, conical, with trenchant posterior margins. In the males of the great Baboons, Orangs, and Chimpanzees, they acquire a size and development rivalling that in the typical Carnivora. Their most formidable development is exemplified in the Maudrill (*Cynocephalus maimon*). In the Mandrill, the upper canines descend behind the crowns of the lower, and along the outside of the first lower premolars, the crowns of which appear bent back by their action. Deeply grooved in front, like the poison fangs of some snakes, they have obtained the name of "*dentes canaliculati*," whilst their posterior margin is very sharp. They are separated by a long interval from the upper incisors, by a shorter one from the first upper premolar. A comparison of the human canines with the same teeth in the Gorilla (see plate I), Chimpanzee, and Orangs, demonstrates one of the most striking

conceived, by supposing an *Incisor*, with its corners rubbed off,

structural differences between Man and those forms of life which anatomically approach him nearest.

In its conical shape, the human cuspidatus does not present an exception to the configuration of canine teeth generally ; but its peculiar characteristics are its small, absolute, and relative size, and the absence of any difference in its development in the two sexes. It is more deeply implanted, and possesses a stronger fang than the incisors ; its external surface is convex, its internal is flat, or sub-concave, and presents a slight prominence at its base. In the genus *Troglodytes*, the canine tooth in both sexes is of greater relative size than in the human subject, but in the male sex it attains vastly superior dimensions. In the adult male of both *Tr. Gorilla* and *Tr. Niger*, the apex of the upper cuspidatus extends a little below the alveolar border of the lower jaw when the mouth is shut : in the female the apex is lodged in the interval between the lower canine and first premolar. The crown of the upper laniary in the male is pointed, and of a conical figure ; its external surface is convex, presenting a longitudinal depression anteriorly ; posteriorly, it is somewhat flatter, and is bounded by a sharp cutting edge ; the internal surface is divided into an anterior and posterior facet by a longitudinal rising ; the posterior is concave, and both it and the anterior are grooved longitudinally. In the male Gorilla the cuspidatus has a more outward direction than in the Chimpanzee ; the anterior internal groove is deeper, and the internal ridge is more developed ; the posterior internal groove is continued lower on the fang, and the posterior trenchant edge is more produced towards the base. The size of the canine in the male Gorilla is greater, as compared with the incisors, than in the Chimpanzee ; in the male of both species, it attains twice the size of the same tooth in the female. In length the enamelled crown of the superior canine in the adult male *Tr. Gorilla*, measures one inch and four lines. An equal sexual distinction is maintained in the dimensions of the canines in the lower jaw. The crown of the lower canine is conical and trihedral, its external surface is convex, and the two internal are flat or slightly concave, and converge to an almost trenchant edge. Anteriorly, a ridge separates the external from the antero-internal surface. The entire length of the lower canine in the male Gorilla, is two inches and a half ; in the Chimpanzee, two inches. The crown in the larger species is one inch and a quarter in length, and nearly an inch in breadth at its base ; in *Tr. Niger*, it is three-quarters of an inch in length, and two-thirds of an inch across the base. In the Orangs (*Pithecus*), a like high degree of development is attained by the canines, which present the same sexual difference, and the same general configuration. The laniary of the Great Orang does not, however, quite equal that of the Gorilla, either in length or breadth. The posterior trenchant margin of the upper tooth is mode-

so as to end in a narrow point, instead of a thin edge; and

rately developed, but the anterior longitudinal groove, which is present in both species of *Troglodytes*, is absent.

In the typical genus of the *Carnivora*—the genus *Felis*—the canines are remarkable for their size and development. In the Lion, for instance, this tooth is of extraordinary strength, deeply implanted by a fang of considerably greater length and thickness than the enamelled crown. The crown is conical and sharp-pointed, slightly recurved, convex in front, with one or two grooves on the outer side, almost flat on the inner side, with a trenchant posterior edge. The canines are proportionately somewhat smaller in the Hyænas than in the *Felidæ*. In the *Viverridæ* they are less developed, and their crowns are almost smooth. In the genus *Canis*, the enamel-pointed crown forms about half the length of the tooth, it is curved, sub-compressed, and the surface is uninterrupted by longitudinal depressions. Canines are well developed in the sub-family of the Badgers (*Melidæ*) : they are well-pointed, with a trenchant, posterior edge. In the Indian Badger, the crown is more compressed than in the European. All the Seals (*Phocidæ*) are provided with large and strong canines. In the great proboscidian and hooded Seals, this tooth attains very formidable development, especially in the male; its fang is curved, thick, and subquadrate. Canines are only present in the upper jaw of the Walrus, but their absence in the lower is fully compensated by the extraordinary size and development attained by the superior pair. These teeth grow to an enormous length; they descend and project from the mouth like tusks, inclining slightly outwards, and bending backwards. A transverse section of the tooth presents an oval, with a shallow, longitudinal groove on the inner side, and one or two narrower, longitudinal impressions on the outer. Its growth being uninterrupted, the base continues widely open. The tusks of the Walrus are not merely weapons of offence and defence, but they are most serviceable in aiding the animal to clamber over blocks of ice. In a large extinct Carnivore, the *Machairodus*, the upper canines attained a development almost rivalling that of the tusks of the Walrus. Their shape, however, as well as the number and conformation of the other teeth, indicate a close relationship to the genus *Felis*. The crown of the Canine in *Machairodus* is more compressed and trenchant than in the *Felidæ*. It passed outside the lower jaw when the mouth was closed, as is indicated by a depression on the outer side of the inferior maxilla, between the canine and first molar. Both the anterior and posterior edges of this long, falciform tooth are finely serrated. The lower canine of *Machairodus* is small, and appears to form the terminal tooth of the incisor series.

Herbivora.—Amongst the vegetable feeders, the presence of canine teeth in one or both jaws is by no means constant. Thus, in the hollow-horned Ruminants, these teeth are invariably absent in the upper jaw. In many, also, of the solid-horned Ruminants (*Cervidæ*) upper canines

the fang differs from that of an *Incisor*, only in being much larger.

The outside of the body of a *Cuspidatus* projects most at

are equally wanting. In those which possess them, they are frequently confined to the male sex, and when present in both sexes, they are always of minor development in the female. Amongst the deer tribe, upper canines reach their largest size in the small musk-deer, especially in the typical species (*Moschus moschiferus*), in the males of which the proportions attained by these teeth are intermediate between those in the *Machairodus* and in the Walrus. It is interesting to notice that, although these teeth are not present in the upper jaw in the more typical *Ruminantia*, that yet they existed in those extinct forms which appear to have been replaced by them. Thus, amongst the fossil remains of the eocene tertiary period, we find canines present in both jaws in the *Dichodon cuspidatus*, an animal which, as respects its dentition, appears to have formed a connecting link between another eocene tertiary form —the *Anoplotherium* and existing *Ruminantia*. In the *Dichodon*, these teeth closely resemble the incisors. Although a little larger, and having a low point, they are yet more trenchant than piercing. The presence and development of upper canines in the Ruminants appear to bear an inverse relationship to the presence and growth of horns. Those ruminating animals which have typical horns are unprovided with these teeth. First present in the periodically hornless deer, they attain their largest dimensions in the absolutely hornless Musks; whilst, in the Camelidæ, the upper canine is not only of formidable dimensions itself, but, as we have already seen, it is accompanied by an upper laniariform incisor. In the Llamas of the New World, the upper canines are more feeble than in the camels and dromedaries. The inferior canines in the ordinary Ruminants are procumbent, and appear to form part of the same series with the lower incisors. They may, however, be distinguished by the lateness of their eruption, and frequently by their form. Thus, in the Musk-deer, they are smaller and more pointed than the incisors; in the Giraffe they have a much larger crown, which is bilobed. In the Camelidæ, they are more erect in position, and present a laniariform configuration. In the *Aucheniæ*, a short diastema separates them from the incisors.

On turning to the non-ruminant, even-toed Ungulates, we find, in the Hog tribe, a remarkable development of the upper canine teeth. The tusks or upper canines of the Wild Boar curve forwards, upwards, and outwards; their sockets having a similar direction, and being strengthened above by a ridge of bone, which is remarkably developed in the African Masked Boar. The enamel on the inferior side of the tusk is ribbed longitudinally; a narrow strip of hard enamel is also laid upon the anterior part, and another upon the posterior concave angle forming the point of the tusk, which is worn obliquely upwards from before, and

the side next the *Incisores*, being there more angular than any where else.

The Enamel covers more of the lateral parts of these Teeth than of the *Incisores*; they stand perpendicularly, or nearly so, projecting farther out in the circle than the others, so that the two *Cuspidati* and the four *Incisores* often stand almost in a straight line, especially in the Lower-Jaw.

This takes place only in adults, and in them only when the second Teeth are rather too large for the arch of the Jaw; for we never find this when the Teeth are at any distance from one

---

backwards from that point. In the female, the tusks are of much smaller dimensions than in the male. Their development in the latter is arrested by castration. A similar tendency to excessive development of the canine teeth is to be observed in the Hippopotamus. In this animal the lower canines are extremely massive and large; the crown is curved and subtrihedral, the angle dividing the two anterior sides, which are convex and enamelled, being rounded off; the posterior side is almost entirely occupied by the oblique surface abraded by the upper canine. The upper canines are trihedral, with a wide and deep longitudinal groove behind; they curve downwards and outwards; the exposed part of the crown, which is worn at the forepart from above obliquely downwards and backwards, is very short. As in the incisors, the base of the canine is simple, and excavated for a large, persistent matrix, which insures the perennial growth of the tooth.

Amongst Perisso-dactyle Ungulates, we find canines present in the genera *Equus* and *Tapirus*, absent in the Rhinoceroses and in the Hyrax. In the Horse, these teeth are small in the stallion, less in the gelding, and rudimental in the mare. The upper canines are placed in the middle of the long interspace which separates the incisors from the premolars. The crown, when unworn, is characterised by the folding in of the anterior and posterior margins of the enamel, which includes an extremely thin layer of dentine. The lower canine, as in *Ruminantia*, is close to the incisors, but is of a more pointed shape. In the Tapir, the canines are of cuspidate form, the crown is much shorter than the root, and does not project beyond the lips. The outer surface of the crown is convex, and is divided by sharp margins from a less convex inner surface. The upper canine is separated from the incisors by a short interval, which receives the crown of the lower. The lower forms part of the same semi-circular series with the inferior incisors. In the Proboscidians there are no canines. (1)]

---

(1) Owen, *op. cit.*

another, or in young subjects. Their points commonly project beyond the horizontal line formed by the row of Teeth, and their fangs run deeper into the Jaws, and are oftener a little bent.

In the Upper-Jaw they are rather longer, and do not project much beyond the circle of the adjacent Teeth; and in this Jaw they are not placed vertically, their bodies being turned a little forwards and outwards.

When the Jaws are closed, the *Cuspidatus* of the Upper-Jaw falls between, and projects a little over the *Cuspidatus* and first *Bicuspis* of the Lower-Jaw. When they are a little worn down by use, they commonly first take an edge somewhat like a worn *Incisor*, and afterwards become rounder. (*h*)

The use of the *Cuspidati* would seem to be, to lay hold of substances, perhaps even living animals; they are not formed for dividing as the *Incisores* are; nor are they fit for grinding. We may trace in these teeth a similarity in shape, situation, and use, from the most imperfectly carnivorous animal, which we believe to be the human species, to the most perfectly carnivorous, viz. the lion. (*i*)

(*h*) [In the different varieties of the human species, the canines present no constant variation. It has been asserted that they are more acuminated in the Melanian races. An examination of a large number of Negro and Negrito crania, however, has led the Editor to the conclusion that there is no constant difference in the configuration of the cuspidati in those races. Their prominence, like that of the incisors, is generally in proportion to the greater or less degree of prognathic conformation.]

(*i*) That our conclusions as to the functions of an organ as it exists in man, when drawn exclusively from analogous structures in the lower animals, will frequently prove erroneous, is strikingly shown in these observations on the use of the cuspidatus. The simple and obvious use of this tooth, in the human species, is to tear such portions of food as are too hard or tough to be divided by the incisors; and we frequently find it even far more developed in animals which are known to be exclusively frugivorous. Not only is its structure wholly unadapted for such an object as that assigned to it in the text, but there is no analogous or other ground for supposing that man was originally constructed for the pursuit and capture of living prey. His naturally erect posi-

OF THE BICUSPIDES.

Immediately behind the *Cuspidati*, in each Jaw, stand two Teeth, commonly called the first and second Grinders, but which, for reasons hinted at above, I shall suppose to constitute a particular class, and call them *Bicuspides*. (*j*)

These (viz. the fourth and fifth Tooth from the symphysis of the Jaw) resemble each other so nearly, that a description of the first will serve for both. The first indeed is frequently the smallest, and has rather the longest fang, having somewhat more of the shape of the *Cuspidatus* than the second.

The body of this Tooth is flattened laterally, answering to the flat side of the fang. It terminates in two points, viz. one external and one internal. The external is the longest and thickest; so that on looking into the mouth from without, this point only can be seen, and the Tooth has very much the appearance of a *Cuspidatus;* especially the first of these Teeth. The internal point is the least, and indeed sometimes so very small, that the Tooth has the greatest resemblance to a *Cuspidatus* in any view. (*k*) At the union of the points the Tooth is thickest, and thence it loses in thickness, from side to side, to the extremity of the fang; so that the fang continues pretty

---

tion, and the structure of the mouth, would render this impossible by the means inferred by Hunter, and the possession of so perfect an instrument as the hand, obviates the necessity of his ever employing any other organ for the purpose of seizing or holding food of whatever description.—T. Bell.

(*j*) [The premolars (*bicuspides* in human anatomy) are those teeth which succeed the deciduous molars (*Vide* Note b., p. 59). For the sake of convenience of description, the principal variations in the form and number of the premolar series, in the different orders of Mammalia, will be noticed in connection with the comparative anatomy of the true molars.]

(*k*) [From the greater size of the outer cusps, as compared with the inner of both premolar teeth, the outer curve formed by the premolar part of the dental arch is greater than the inner. In the upper bicuspid, the outer and larger cusp is divided from the inner by a deep, straight fissure or groove; in the lower bicuspid, this fissure describes

broad to the point, and is often forked there. (*l*) All the Teeth hitherto described often have their points bent, and more particularly the *Cuspidati*.

The Enamel passes somewhat farther down externally and upon the inside, than laterally; but this difference is not so considerable as in the *Incisores*, and *Cuspidati*; in some indeed it terminates equally all round the Tooth. They stand almost

---

a curve concave towards the outer cusp. In some cases the curved groove in the lower premolar is crossed by a ridge, which extends from the outer to the inner cusp. The inner cusp in the lower tooth is less developed than the same tubercle in the upper. (1)]

(*l*) [The fang of the lower premolar is single, long, sub-compressed, and conical. In the upper jaw, although at first sight the fang of the premolar might be supposed to be single, an examination of the pulp cavity, which is bifurcated, shows that it really consists of two connate fangs, an internal and external, which generally, by a groove on the surface and by a bifid extremity, evince a tendency to separation. (2) More rarely, the external and internal fangs are completely separated; according to Mr Nasmyth, this conformation is generally confined to the anterior upper bicuspid. (3) In some rare and exceptional instances the Editor has seen the external division of the root of the upper bicuspid again divided into an anterior and posterior fang, making an implantation by three fangs—the normal implantation of these teeth in the Anthropoid Apes. Mr Tomes, in his Lectures on Dental Physiology and Surgery, notices this conformation, and instances a Chinese skull in which the upper anterior bicuspid has an implantation by three fangs, like a molar. (4) Such an anomaly, however, is peculiar to no race of Man. The Editor has noticed it in the cranium of a Frenchman, in a North American aboriginal, and in two crania of slaves from the banks of the Mississippi. He has also seen the division indicated by a deep groove on the external fang of an upper premolar in the cranium of a Hindoo. In all the instances, save one, which have fallen under his notice, it was the anterior upper premolar which displayed this complex implantation. In the exceptional case, the left posterior upper premolar was implanted by three fangs, the right by two.

---

(1) Owen, op. cit.

(2) Ibid.

(3) Nasmyth, Researches on the Teeth, p. 68.

(4) Tomes, Lectures on Dental Physiology and Surgery, p. 13.

perpendicularly, but seem to be a little turned inwards, especially the last of them.

In the Upper-Jaw they are rather thicker than in the Lower, and are turned a very little forwards and outwards. The first in the Upper-Jaw falls between the two in the Lower. The second falls between the second and the first Grinder: and both project over those of the Lower-Jaw, but less than the *Incisores* and *Cuspidati*. (m)

The *Bicuspides*, and especially the second of them, in both Jaws, are oftener naturally wanting than any of the Teeth, except the *Dentes Sapientiæ;* thence we might conjecture that they are less useful; and this conjecture appears less improbable, when we consider, that in their use they are of a middle nature between Cutters and Grinders; and that in most animals, so far as I have observed, there is a vacant space between the Cutters and Grinders. (n) I have also seen a Jaw in which the

---

(m) [The premolars present no variation in configuration in the different varieties of Man. Like the molars, as a rule, they are of large size in the Melanian races, especially in the Australian; although, even among these, exceptions occur in which these teeth are not of greater dimensions than in some individuals of other races.

(n) In many instances in which the second bicuspid (p. 4) appears to be absent, it is prevented coming into place by the retention of the second deciduous molar to a late period of life. The retention of the first deciduous molar may, in like manner, prevent the appearance of the first bicuspid. It would appear, however, that this is more rare. It occasionally also happens that the second bicuspis is altogether undeveloped, in which case the second deciduous molar permanently supplies its place. Mr. Tomes states that, according to his experience, the lateral incisors are more frequently absent than any other teeth, with, perhaps, the exception of the wisdom.(1) Comparative anatomy does not warrant us in seeking any analogy between the occasional absence of the second bicuspis in Man, and the existence of a *diastema* between the grinders and cutters in the lower animals, for the second bicuspis is the homologue of the tooth (p. 4), which of all the premolar series is most constant in placental diphyodonts. The premolars in the higher Mammalian dentition form part of the same series with the molars, in front of which is the *diastema*, separating them from the canines and incisors; and reduction in the number of premolars always takes place by the suppression of one or more of the anterior teeth.

---

(1) Tomes's System of Dental Surgery, p. 221.

first *Bicuspis* was of the same shape and size as a Grinder, and projected, for want of room, between the *Cuspidatus* and second *Bicuspis*. These and the Grinders alter very little in shape on their grinding surfaces by use; their points only wear down, and become obtuse.

OF THE GRINDERS.

In describing the Grinders (*o*) we shall first consider the first and second conjunctly, because they are nearly the same in every particular; and then give an account of the third or last Grinder, which differs from them in some circumstances.

The two first Grinders differ from the Bicuspides, principally in being much larger, and in having more points upon their body, and more fangs.

(*o*) [The following is a brief sketch of some of the more striking modifications presented by the molar and premolar series in the Mammalian orders. It has been compiled from the previously-quoted work of Professor Owen.

In the order *Bruta*, the teeth for the most part belong to the molar series. The characteristics of the dentition in this order are, that each tooth is implanted by an undivided base, the implanted portion continuing of the same thickness as the crown. In no species of this order possessing teeth are these organs implanted by divided roots, or is there a cervix separating the crown of the tooth from the implanted part. The base is excavated for the entrance of a persistent pulp, and the tooth continues to grow throughout life. The teeth are devoid of a coating of enamel. In the Cape Ant-eater (*Orycteropus Capensis*), the teeth all belong to the molar series. Its dental formula is $\frac{5}{5}$: $\frac{5}{5}$:. The two anterior teeth in the upper jaw, and the anterior tooth in the lower jaw, are very small, and not unfrequently concealed by the gum, or wanting, especially the first, in the upper jaw. The tooth in front of the penultimate, and the penultimate are the largest. These two teeth have a depression on their internal and external sides, which gives their transverse section somewhat of an hour-glass form, each tooth is implanted by a truncated undivided base. The tooth of the Orycteropus consists of long, slender, prismatic denticles of dentine cemented together; the base of each denticle presents the conical opening of a persistent pulp cavity. These openings give a porous character to the base of the entire tooth. The aperture at the base of each denticle leads to a canal from whence the dentinal tubules radiate. The denticles are united laterally by cement; they slightly decrease in diameter, and occasionally bifurcate as they approach the grinding surface of the

The body forms almost a square, with rounded angles. The grinding surface has commonly five points, or protuberances,

---

tooth. The effects of attrition on that surface are compensated by the continued growth of the denticles. In the Armadillos (*Dasypus*) the dental organs attain the greatest degree of hardness found in the order. Each tooth consists of a small central axis of vascular dentine, surrounded by a body of ordinary dentine, which forms the principal bulk of the tooth, and the external surface of which is covered by a thin layer of cement. The teeth in the great Armadillo (*Priodon gigas*), amounting to from 94 to 100 in number, are all to be classed as molars. They are of small size, simple in form, and they increase, especially in breadth, from before backwards. The implanted portion of the tooth is as thick as the uncovered part, and is widely excavated for a persistent pulp. The dentition in the great extinct Armadillo (*Glyptodon*), presents a conformation more complex, and better adapted for vegetable feeding than that of the existing species. The tooth of the Glyptodon is long and rootless, as in the existing Armadillos; it is compressed laterally, and deeply indented on each side by two angular longitudinal and opposite grooves, which give the tooth a three-lobed configuration. The centre of this tooth consists of osteo-dentine, which is here harder than the dentine or the cement. It consequently forms a ridge on the grinding surface of the tooth, which extends along the middle of its long axis and sends a prolongation on each side, into each of the three rhomboidal lobes into which the surface is divided. In the phyllophagous Bruta—the sloths—the teeth which are few in number, are composed of a large central axis of vaso-dentine, surrounded by a thin investment of hard dentine, and a thicker covering of cement. They possess the characters of uninterrupted growth, and implantation by a simple deeply excavated base, not separated from the exposed portion by a cervix. Such were the teeth of the extinct Megatherium, the most gigantic of the sloth tribe. Its dental formula was $\frac{4}{4} : \frac{4}{4} :$. The teeth of the Megatherium present a more or less tetragonal figure, and the grinding surface of each is traversed by two transverse angular ridges. The sloping side of the ridge nearest the centre of the tooth is formed by vascular dentine, the side farthest from the centre by cement, whilst the hardest material—the unvascular dentine—forms the summit of the ridge.

In those of the true *Cetacea*, which possess true functional teeth—*e. g.*, the Cachalot, the Dolphins, and Porpoises—these organs, with the single exception of the common Dolphin before referred to—(note *d.*, p. 61), are limited to the maxillary and mandibular bones. They, may, therefore, be briefly noticed here. In the Cachalot (*Physeter macrocephalus*), the visible dentition is confined to the lower jaw: it consists, in the male, on each side, of about twenty-seven sub-incurved, conical, or ovoid teeth. In the female, the teeth are fewer—twenty-three in each

two of which are on the inner, and three on the outer part of the Tooth; and generally some smaller points at the roots of these larger protuberances. These protuberances make an irregular cavity in the middle of the Tooth. The three outer points

ramus—and shorter than in the male. The smallest teeth are those at each extremity of the series. They are implanted by fangs, and are lodged in a wide and moderately-deep groove in the jaw, which is imperfectly divided into sockets, the septa reaching only about half way from the bottom of the groove. The intervals separating these teeth are about as broad as themselves. The sockets are too wide and too shallow to retain the teeth independently of the soft parts; hence, in the dried condition, when the ligamentous gum is stripped off, it often happens that the whole row of teeth is brought away with it. These teeth, when the mouth is closed, enter depressions in the upper gum. A few rudimentary primitive teeth (representing those which even in the true Whales (*Balænidæ*), are always present in the jaws in the fœtal condition) are retained in the upper gum in the Cachalot. They are more curved than the functional teeth in the lower jaw. In the young Cachalot, the tooth is tipped with enamel, but after a time the apex becomes worn down, and the tooth consists of a hollow cone of dentine coated by cement, and filled more or less with ossified pulp. In the Dolphins teeth are present in both jaws; they vary considerably in number in different species. The teeth are generally conical, strong, and pointed, and implanted by a single fang, but more firmly than in the Cachalot. In most Dolphins, the anterior teeth become obtuse by the wearing down of their sharp apices; whilst the posterior teeth continue sharp-pointed; the usual position of piercing and bruising teeth being thereby reversed. An exception to this, however, is presented by the Gangetic Dolphin (*Platanista gangetica*). In it the anterior teeth retain their prehensile characters, whilst the crowns of the posterior become worn away to the base. Professor Owen has noticed that sometimes the base of the posterior tooth in the Gangetic Dolphin, is divided into two short fangs, a conformation probably unique in existing carnivorous Cetacea. A large extinct animal, which, from the characters and microscopic structure of its teeth, Professor Owen believes to have been a Mammal, allied to the Cetaceans, and which he has named *Zeuglodon*, or yoke-tooth, had its posterior teeth implanted by double fangs. The crown of the posterior tooth in the Zeuglodon is contracted from side to side in the middle of its base, so as to give its transverse section an hour-glass form; and the deep longitudinal opposite grooves which produce this form become deeper and deeper as they approach the socket, and at length unite to divide the root into two fangs. The upper part of the crown has its anterior and posterior margins strongly serrated. The anterior teeth of the Zeuglodon have

do not stand so near the outer edge of the Tooth, as the inner do on the inside; so that the body of the Tooth swells out more from the points, or is more convex, on the outside. The body towards its neck becomes but very little smaller, and

single roots; they are conical, sharp, slightly recurved, and compressed laterally. A fragment of the under jaw of the Zeuglodon has been figured by Dr Carus, in which a worn-out deciduous molar, is apparently displaced and succeeded vertically by a premolar. This would indicate an affinity to the *Sirenia* rather than to the true Cetacea. For the latter, like the Bruta, are monophyodonts.

In the Aquatic Pachyderms (*Sirenia*) known also as Herbivorous Cetacea, we find molars distinguished by configuration and position. The number of these teeth in the Australian Dugong does not exceed twenty-four. In the Malayan species, there are not more than twenty— viz., five on each side in each jaw. But the entire series is never in use together, the first is shed before the last has cut the gum. They increase in size from before backwards. The four first teeth are more or less cylindrical in figure; each is implanted by a single fang, which, in its turn, becomes completed and solidified by the contraction and obliteration of the basal cavity. The fifth or hindmost molar is more complex. A transverse section of its crown presents an approach to the hour-glass configuration of the posterior teeth in the Zeuglodon. The root shows a tendency to division, whilst the pulp is maintained in a wide basal pulp cavity, to supply the waste of the crown. In the American Manatee the formula of the molar series is $\frac{1.9}{9} \cdot \frac{1.9}{9} = 38$, but these teeth are never all in place and use at the same time. In the upper jaw the first two are shed before the last two appear; in the lower, seven are usually in use simultaneously. The first molar, in both jaws, is small and simple. Beyond the second, the crowns of the upper series are square; the grinding surface supports two transverse ridges, each of which is surmounted by three tubercles; it is also bounded by an anterior and posterior basal ridge. Each upper molar is implanted by three diverging roots, two on the outer, and one on the inner side. The crowns of the four or five anterior lower molars resemble those above; the remainder present a large posterior tubercle; they are implanted by two fangs, which enlarge towards the extremity, where they bifurcate. The grinders of the Manatee not succeeding deciduous teeth, are all referable to the true molar series. They consist of dentine, with a general investment of cement, and a coronal coating of enamel. The large number of the molar teeth, and the simple conical character of the first tooth which is separated by an interval from the two-ridged molars, are indications of affinity to the cetaceous character, whilst the shedding of the anterior teeth before the development of the posterior —a circumstance presenting an analogy to the mode of succession

there divides into two flat fangs, one forwards, the other backwards, with their edges turned outwards and inwards, and their sides consequently forwards and backwards: the fangs are but very little narrower at their ends, which are pretty broad, and

---

of the molar teeth in the Elephant—together with the shape, structure and implantation of these organs indicate a near approach to the pachydermal type. The molars of the Dugong resemble in the double-ridged character of the crowns those of the Tapir, Dinotherium, Diprotodon, and Kangaroo.

The typical number of the molar teeth in the Marsupial Order is seven on each side of both jaws; of which the three first succeed three deciduous teeth, and are premolars, whilst the true molars are four in number. In the Thylacine, there are three premolars and four molars on each side, both above and below. The premolar has a simple compressed conical crown, with a posterior tubercle, which is most developed in the third. The crowns of the upper molars are of an irregularly triangular figure. The outer part of the crown is raised into three cusps, of which the middle is the largest, and from the inner side of the crown projects a small, strong, obtuse lobe. The lower molars are compressed and tricuspid; the centre cusp being the longest, especially in the third and fourth, which resemble closely the dents carnassières of the *Felidæ*. In the Dasyures, the premolars are reduced to $\frac{2}{2} \cdot \frac{2}{2}$; the molars are $\frac{4}{4} \cdot \frac{4}{4}$. The premolars have simple crowns. The crowns of the upper true molars are triangular; the first has four sharp cusps; the second and third each five, whilst the fourth, which is smallest, is tricuspid. The last lower molar is nearly of equal size with the third; it presents four cusps, of which the outer is the largest; the second and third have five cusps, three on the inner side, and two on the outer; the first is quadricuspid. An extinct carnivorous Australian Marsupial, to which Professor Owen has given the name of *Thylacoleo*, had a true carnassial tooth, upwards of an inch and a half in fore and aft extent, and one inch in height. In the lower jaw, the crown of this tooth consists entirely of blade; in the upper, there is a feeble tubercle superadded. In the smaller species of carnivorous Marsupials, the character of the molars approaches the insectivorous type, in being more cuspidate than sectorial. This is the case in the *Phascogale*, in which the formula of the molar series is $p. \frac{3}{3} \cdot \frac{3}{3}$. $m \frac{4}{4} \cdot \frac{4}{4}$. In *Myrmecobius*, with the same number of premolars, the true molars are increased to $\frac{5}{5} \cdot \frac{5}{5}$. The molars in *Myrmecobius* are multicuspid; they are separated from each other, as are also the premolars, by short intervals; both molars and premolars are implanted by two fangs, an implantation general in the Marsupial Order. The lower molars are directed obliquely inwards, so that their outer surfaces come into contact with the working surfaces of the upper. The premolar is compressed and triangular, with the apex

L

often bifurcated. There are two cavities in each fang, one towards each edge, leading to the general cavity in the body of the Tooth. These two cavities are formed by the meeting of

---

somewhat recurved, and the base obscurely notched before and behind. The extinct *Phascolotherium* had three premolars and four molars in the lower jaw; the first premolar and fourth molar are the smallest. The five larger teeth are each tricuspid, the middle cusp being the largest; but the crown is girt by a basal ridge, which projects a little beyond the anterior and posterior cusps, and gives the tooth a quinque-cuspid character. The dentition of the genus *Amphitherium* resembles, in many respects, that just described; but not only are the molars increased in number, as in *Myrmecobius*, but the premolars are also six on each side in the lower jaw. The dentition of the upper jaw in *Phascolotherium* and *Amphitherium* is at present unknown, but it probably corresponded with that of the lower. In the Opossum (*Didelphys*), the formula of the molar series is $p. \frac{3}{3}: \frac{3}{3}: m. \frac{4}{4}: \frac{4}{4}:$. The molars are tuberculous in structure, and depart more or less from the type of the true flesh-feeders. In *Phascolarctos*, *Hypsiprymnus*, *Macropus*, and *Phascolomys*, the premolars are reduced to $\frac{1}{1}: \frac{1}{1}:$, whilst the molars are $\frac{4}{4}: \frac{4}{4}:$. In the Koala (*Phascolarctos cinereus*) the premolar is compressed, and has a cutting edge; the molars are quadricuspid, with the rudiment of a "cingulum." In the Potoroos, or Kangaroo Rats (*Hypsiprymnus*), the single premolar is of large size, and has a peculiar trenchant form: in the young tooth, the crown is indented, especially on the outer side, by a series of vertical grooves. In the Arboreal Potoroos of New Guinea, the premolar nearly equals in fore and aft extent, the three succeeding molars. The true molars are surmounted by four trihedral, pyramidal cusps; the internal angles of opposite cusps are continued into each other across the grinding surface, giving rise to two transverse ridges. In the old tooth, the cusps and ridges become worn down. In the Kangaroos (*Macropus*), the crowns of the true molars present two transverse ridges, with a broad anterior talon, and a narrow hinder one. In most species, a spur connects the anterior with the posterior ridge, and another the anterior ridge with the front talon. The molars in the gigantic, extinct Diprotodon were five in number on each side in both jaws; they presented the same double, transversely-ridged conformation, with the addition of an anterior and posterior low basal ridge. The dentition in the Wombat (*Phascolomys*) is peculiar amongst Marsupialia. The premolar has lost the compressed character it has in the Koala and Kangaroos. It presents a wide, oval, transverse section: the superior premolar tooth is traversed on the inner side by a longitudinal groove. The superior molars are each also divided by a wide, internal, longitudinal groove, into two prismatic lobes, one angle of each prism being directed inwards. The inferior molars are, in like manner, divided into

the sides of the fang in the middle, thereby dividing the broad and flat cavity into two; and all along the outside of these (and all the other flat fangs) there is a corresponding longitu-

---

two three-sided lobes, by an external groove. All the molars are curved —in the upper jaw, the concavity of the curve is directed outwards; in the lower, inwards. The premolar is about half the size of a true molar. Like the incisors, the premolars and molars of the Wombat have persistent pulps; they are therefore implanted by an undivided base. In this respect, these teeth differ from the molars of all the other Marsupials, whilst they resemble those of the dentigerous Bruta and herbivorous Rodents.

The crowns of the molars in the order *Insectivora* are always broader in the upper jaw than in the lower; they also always present several sharp points or cusps upon the working surface. The dentition of the *Chrysochlore*, or iridescent Mole of the Cape, is remarkable for the number of the molar teeth, for their shape, and for their separation by vacant intervals, as in many Reptiles. The number of the molar series is $p. \frac{1}{2} : \frac{1}{2} m. \frac{9}{5} : \frac{9}{5}$. This formula is based on the shape of the teeth, as the facts relating to their vertical displacement and succession are not known. The premolar in the upper jaw has a simple, compressed, tricuspid crown. The crowns of the true molars are thin plates, compressed antero-posteriorly, with two notches on the working edge, and a longitudinal groove along the outer and thicker margin. They are separated by intervals. In the lower jaw, the molars attain a much greater height above the sockets; they are separated by wider intervals than those above, and the two series interlock when the mouth is closed. In mastication, the anterior margin of the lower tooth works against the posterior margin of the opposed upper tooth. The crowns of the lower molars are long plates, compressed antero-posteriorly; the inner margin is the thickest, and is surmounted by two points. The outer margin terminates in a single point, which is most prominent. In an extinct small insectivore, the *Spalacotherium*, there were ten teeth in the molar series on each side of the lower jaw; of these, at least six had tricuspid crowns, and bore a marked resemblance to the same teeth in the *Chrysochlore*. In the Shrew Moles of America (*Scalops*) the number of the molar series is reduced to $p \frac{3}{3} : \frac{4}{4}; m \frac{4}{4} : \frac{3}{3}$. The restriction of the true molar characters to the last three teeth on each side in each jaw, is an important step towards the typical Mammalian dentition. Of the premolars in the upper jaw, the last is the largest, and has a trihedral, pointed crown. The true molars have large crowns; each supports six cusps, four on the outer and two on the inner side. In the lower jaw, the long crown of each true molar is divided into two parallel three sided prisms, each of which is surmounted by three points; each prism has one of its angles turned outwards, and one of the faces inwards: the interspaces

dinal groove. These fangs at their middle, are generally bent a little backwards.

between the angles give a grooved appearance to the outer surface of the molar series in *Scalops*. In the common Mole (*Talpa Europœa*), the normal number of teeth is retained : there are four premolars and three true molars on each side, both above and below. In the upper jaw, the three first premolars have small, simple, conical crowns, and are implanted by two fangs ; the fourth premolar has a large trihedral crown, and three fangs. The true molars are multicuspid. The middle one is the largest— it is quinque-cuspid, with usually four fangs ; the third, which is the smallest, has a tricuspid crown and three fangs. In the lower jaw, the first premolar is the largest. Each lower premolar is implanted by two fangs, and has a small posterior talon at the base of the conical crown. The lower true molars, of which the middle is the largest, are quinque-cuspid, and implanted by two fangs. In the *Solenodon* of Hayti, a species allied to the Shrews (*Sorex*) and Water-moles (*Mygale*), seven teeth succeed the incisors in both jaws. Of these, in the upper jaw, the two anterior have trihedral conical crowns ; the five succeeding present, in addition, an external, tuberculate basal ridge. In the lower jaw, the four last teeth have multicuspid crowns. The typical Shrews (*Sorex*) have two small, and three large, multicuspid molars in the lower jaw : in the upper, the number of premolars is variable. The last true molar is usually of small size. The differences in the number of teeth in the upper jaw, together with the variations in the shape of the large incisors, are points which serve to characterise the sub-genera of Shrews. The formula of the molar series in the Tupaias (*Glisorex*) is $p\frac{3}{3}.\frac{3}{3}.\ m\frac{3}{3}.\frac{3}{3}.$. The premolars increase in size and complexity as they approach the molars : the two first of the latter in the upper jaw have each six cusps, the corresponding teeth in the lower jaw are quinque-cuspid. The last true molar, both above and below, is the smallest, and has three cusps. The Gymnure is one of the rare instances among existing Mammals in which the typical dental formula is retained. It has four premolars and three true molars on each side in both jaws. In the upper jaw, the first three premolars have simple crowns—the last is large and quadricuspid. The first and second upper true molars present square and multicuspid crowns. The third is smallest and triangular. In the lower jaw, the third and fourth premolars are much larger than the first and second ; the fourth is tricuspid. In the European Hedgehog (*Erinaceus Europæus*) there are four superior premolars on each side, and two inferior. The true molars are three on each side, both above and below. The last premolar in both jaws is the largest : above, it is quadricuspid, with three fangs ; below, it has a sub-compressed tricuspid crown, with two fangs. Both upper and lower true molars decrease in size from the first to the third. The first and second above are quadricuspid, the corresponding teeth below are narrower and quinque-cuspid. The tropical

The Enamel covers the bodies of these Teeth pretty equally all round.

The first Grinder is somewhat larger and stronger than the

Hedgehogs (*Echinops* and *Ericulus*) resemble the Chrysochlore in the more simple character of the molars, which are compressed antero-posteriorly. Each upper molar presents two outer cusps and one inner, each lower molar one outer and two inner. In both sub-genera, the premolars ($\frac{2}{2}$: $\frac{2}{2}$:) have simple crowns. The molars in *Echinops* are $\frac{4}{4}$: $\frac{4}{4}$:, in *Ericulus* $\frac{5}{5}$: $\frac{4}{4}$:. The Tenrecs or Tailless Hedgehogs of Madagascar, resemble, in the general character of the molar dentition, the last-named genera. The first premolar above is compressed, unicuspid, with a hinder talon, and implanted by two fangs; the second is tricuspid, with three fangs. The three first true molars are tricuspid and three-fanged; the last, which is smaller, has two fangs. In the lower jaw, all the molars have two fangs. It is to be noticed, that the coating of enamel on the teeth of these small insectivores is of great thickness when compared with the body of dentine, a provision for the durability of that cuspidate conformation, which is admirably adapted for cracking and crushing the hard or tough coverings of their insect prey.

Order *Cheiroptera*. The molar series in the bats consists of premolars and true molars, it never exceeds $\frac{3}{3}$: $\frac{3}{3}$:. The true molars are generally bristled with sharp cusps. In the large frugivorous bats (*Pteropus*), the true molars have flat crowns. True molars are absent in the Suctorial or Vampire Bats (*Desmodus*). Two premolars are present on each side in the upper jaw: they are very small teeth; each has a simple, compressed, conical crown, with one fang. In the lower jaw, there are three premolars on each side; the two first resemble the upper, but the third has a larger, compressed, bilobed crown, and is implanted by two fangs.

In the *Rodentia*, the molars are few in number, obliquely placed, the series on each side converging anteriorly in both jaws, and obliquely worn. The variations in the form and structure of these teeth in this order are so numerous as to typify almost all the modifications which they present in the omnivorous and herbivorous genera of other orders. In some Rodents, which subsist on mixed food, and display, as in the case of the true Rat, some carnivorous tendencies—or which feed on the softer and more nutritious productions of vegetables, as the oily kernels of nuts, &c.—the crowns of these teeth are not subjected to that amount of wearing down which requires a prolonged or unlimited growth of dental tissue to replace the lost material. Hence, in some genera, as the Rat, the Porcupine, no more dental tissue is organised after the crown is formed than is necessary to fix it firmly in the jaw. These teeth, therefore, soon acquire roots of ordinary proportions. In a second category of the order, we find the molars with short roots tardily developed, like the same teeth in the Horse and the Elephant. These are they

second; it is turned a little more inwards than the adjacent *Bicuspides*, but not so much as the second Grinder. Both of them have generally shorter fangs than the *Bicuspides*.

---

which subsist on the coarser and less nutritious kinds of vegetable tissues, and whose molars, consequently, are subject to more rapid wear. Such are the Beaver, the Agouti, &c. In them, for a considerable period, the wear of the crown is compensated for by the continued reproduction of the formative matrix, and during this time, the tooth is implanted by a simple, undivided continuation of the crown. When the formative force becomes exhausted, the matrix is simplified by the suppression of the enamel organ, and the dentinal pulp continues to be reproduced only at certain points of the base of the crown, which, by their elongation, constitute the fangs. In a third category, which contains the Cavies, the Capybara, &c., the molars, like the incisors, are perpetually growing, and rootless. These teeth are always more or less curved, by which conformation, as in the case of the incisors, the effects of pressure, transmitted from the grinding surface, on the soft and delicate growing structures at the base of the tooth are obviated. It is in the rootless molars of the strictly herbivorous Rodents that the complexity of the crown is greatest, and that the largest proportion of enamel and cement is interblended with the dentine. In the omnivorous Rats and Mice, the tuberculate molars are almost as simple as the corresponding teeth in the Bears or in Man, which they may be said to typify. In the herbivorous Rodents, we find folds or islands of enamel interposed between the layers of cement and dentine, and, as was pointed out by Baron Cuvier, they have a general transverse arrangement across the crown of the tooth, the opposite to the direction of the temporo-maxillary articulation. The different arrangements of the enamel, dentine, and cement, in the molars of different genera of Rodents, present interesting analogies to the modifications which the arrangements of the same substances undergo in the molars of the different species of Elephant, the Rhinoceros, the Hippopotamus, Ruminants, &c. The molars, although not numerous, vary in number in the different genera. In the Hares and Rabbits, the formula is $\frac{6}{5} \cdot \frac{6}{5}$; in the Pika (*Lagomys*), $\frac{5}{5} \cdot \frac{5}{5}$; in the Squirrels, $\frac{5}{4} \cdot \frac{5}{4}$; in the Dormice, Porcupines, Spring-Rats, Octodonts, Chinchillas, and Cavies, $\frac{4}{4} \cdot \frac{4}{4}$; in the Rats, $\frac{3}{3} \cdot \frac{3}{3}$; in the Australian Water-Rat, $\frac{2}{2} \cdot \frac{2}{2}$; making, with the two incisors in each jaw, twelve teeth, the smallest number known in any Rodent. When the teeth of the molar series exceed three in number, those anterior to the posterior three are premolars, having displaced deciduous molars, and come into place after the true molars, at least after the first and second, even when the deciduous teeth are shed *in utero*.

Order *Quadrumana*. In the genus *Cheiromys*, which manifests a dis-

There is a greater difference between these Grinders in the
Upper and Lower-Jaw, than any of the other Teeth. In the
Upper-Jaw they are rather rhomboidal, than square in their

tinct analogy to the Rodent order in the configuration of the incisors,
there are four teeth in the molar series on each side above, and three
below; they are implanted vertically, and in parallel lines. The molars
are simple in structure, coated with enamel, and they have a flat sub-
elliptic, grinding surface. In the upper jaw, the first is the smallest,
and the second the largest, in the series. The first and last have each
one root, the second and third each three roots. In the lower jaw, the
inequality in size is less than in the upper, but the third is the smallest.
The first lower molar has two roots, the second and third have each one.
The molar series in the genera *Stenops* and *Lemur* consists of six teeth
on each side in both jaws. The three anterior are premolars. In the
true Lemurs, the premolars have the outer part of the crown produced
into a pointed lobe ; the inner forms a tubercle, which is largest in the
second and third. The true molars have the inner division of the
crown so increased as to give it a quadrate form ; the outer division is
divided into two pointed lobes. The first true molar is the largest,
above and below. The Monkeys of the New World (*Platyrhynæ*) manifest their affinity to the Lemurs in the number of the premolars, which
is three on each side of both jaws. The small Marmozets have the true
molar series on each side, above and below, reduced to two.

The Catarhine Quadrumana of the Old World have the same number
of premolars and molars as Man. We shall again best illustrate the
peculiarities of the Human molar dentition, by comparing it with the
grinding series in the highest Apes. The characteristics of the Human
premolars have been already described (*vide* Notes *k* and *l*, pp. 78 and
79). If we contrast with the human premolars the same teeth in the
genus *Troglodytes*, we notice that, in the upper jaw, the premolars
of the Gorilla and Chimpanzee are bicuspid, as in Man ; but they are
larger than the human, and they differ in several other important
particulars. The external cusp of the first premolar, in both species of
Troglodytes, is larger than the inner ; in the second premolar, the inner
is larger than the outer ; in Man, the external is the larger in both.
This alternation in the size of the cusps of the premolars of the Chimpanzees corresponds with, and contributes to, the straight line formed
by the whole grinding series ; whilst the great size of the external cusp
gives to the first premolar, when viewed from without, the appearance
of being greatly superior in size to the second. Each premolar in
Troglodytes is implanted, like the true molars, by three divergent fangs,
two external and one internal, which at their ends curves towards each
other. Professor Owen states that, in one female skull of *Tr. niger*
which he examined, the two external fangs of the second premolar are

body, with one sharp angle turned forwards and outwards, the other backwards and inwards; besides they have three fangs, which diverge, and terminate each in a point; these are almost

connate; and, in a specimen in the Museum of the Great Northern Hospital, which has fallen under the Editor's observation, a similar peculiarity is to be noticed; in these cases, the first premolar presented the usual implantation by three fangs. In the lower jaw in these great Apes, the first premolar, when viewed externally, appears much larger than the second; in the Chimpanzee it is twice the size, in the Gorilla three times the size, of the same tooth in Man. Its crown is somewhat of a three-sided figure, the anterior and outer angle being produced forwards, and making an approach in this respect to the peculiar characteristic of the first premolar in the Baboons. It is surmounted by two trihedral cusps; of these, the outer is the larger and higher, whilst the inner is feebly developed on a ridge extending transversely from the outer cusp. A thick basal ridge belts the inner and posterior part of the crown. The second lower premolar is of a four-sided figure. Its two cusps are placed on the anterior half of the upper surface of the crown; they are united by a transverse ridge, and a third smaller cusp is developed on the inner and posterior angle. Each lower premolar is implanted by an anterior and posterior fang; of these, the former is the larger. They are divergent, and compressed antero-posteriorly. In the Orangs (*Pithecus*), we recognise the same complex implantation of the premolars, by three fangs in the upper, and by two in the lower jaw. The first premolar in the upper jaw differs from the same tooth in the Gorilla and the Chimpanzee, in having the anterior external angle of the base produced, an approximation to the configuration observed in the lower Quadrumana. The outer lobe of this tooth is a little larger than the inner one; in the second upper premolar, both lobes are equal. The two cusps of the first are also more developed than the two cusps of the second. In the lower jaw, the outer cusp of the first premolar attains by far the greater dimensions; from it three ridges proceed—one to the front part of the grinding surface, one to the back part, and a third transversely inwards, where it developes a small tubercle. In the second lower premolar, the inner tubercle attains almost an equal size with the outer. Amongst the distinguishing features of the true molar teeth in Man, are their large size in proportion to the jaw, and in proportion to the other teeth, and the rounded contour of their grinding surfaces. In the upper jaw, the two first teeth of the molar series are quadricuspid. Of the four cusps, the antero-internal one is the largest; and it is connected with the postero-external by a low ridge, on either side of which is a deep groove, extending on the outer side to the middle of the outer surface, on the inner side to the inner surface. The connection of the inner anterior and the outer posterior cusps by the

round, and have but one cavity. Two of them are placed near each other perpendicularly, over the outside of the Tooth; and the other, which generally is the largest, stands at a greater

oblique ridge just mentioned, gives a sigmoid character to the eminences on the grinding surfaces of these teeth. In the last true molar, which is the smallest of the series, the two inner tubercles are blended together; and, in many instances, a groove extends at right angles from the one separating the two outer cusps to the middle of the posterior border of the grinding surface. The upper molars in Man are implanted by three diverging fangs, two external and one internal; but not unfrequently, in the second molar, the two external are found parallel, and occasionally connate. In the third upper molar, the two external fangs are more commonly connate, and sometimes, also, the inner fang is blended with them. Owing to the slow accession of maturity in Man, and the long interval which elapses between the acquirement of the first and last true molars, the former tooth is found more worn in proportion to the other teeth of the same series, than in the Chimpanzees and Orangs. In the lower jaw, the human molar is quinque-cuspid, the fifth tubercle being developed posteriorly, and connected with the postero-external cusp. The fifth cusp is, however, frequently absent in the second tooth of the series, and is most developed in the *dens sapientiæ*. A crucial depression separates and defines the four principal cusps, and, by a bifurcation of its posterior branch, includes the fifth. This bifurcation is most apparent in the third molar. As in the upper jaw, the last named tooth is the smallest of the true grinders. Each molar is implanted by an anterior and a posterior sub-compressed fang, which are grooved along their opposed side. It is not uncommon to find these fangs more or less connate in the second and third teeth of the series. On turning to the genus *Troglodytes*, the first thing which attracts notice is the straight line which the grinding series forms in the Chimpanzee and Gorilla, contrasting strongly with the well-marked curve which the molar teeth describe in the human subject. In the upper jaw, indeed, in these great Apes, the line of grinders makes a slight inclination in the opposite direction to that described by them in Man. Another difference is the smaller relative size of the grinders as compared with the incisors. In this last particular, the Gorilla makes a nearer approach to Man than the Chimpanzee; for in the former the molars are larger relatively, when compared with the incisors, than in *Tr. niger*. In the upper jaw, the first two molars, both in *Tr. Gorilla* and *Tr. niger*, correspond with the same teeth in the human subject, in being quadricuspid, and in having the inner anterior and outer posterior tubercles connected by a low ridge. The sigmoid character thus given to the eminences on the unworn surfaces of these teeth, is highly indicative of the high position in relation to Man occupied by the Chimpanzees above the Orangs and

M

distance on the inside of the Tooth, slanting inwards. In this Jaw these two Grinders are inclined outwards, and a little forwards; they project a little over the corresponding Teeth of

lower Quadrumana, in whom the grinding cusps rise from the surface of the tooth distinct and unconnected. The third upper molar in *Tr. niger* is the smallest of the series, in consequence of the minor development of the two posterior cusps ; also, the oblique connecting ridge between the antero-internal and postero-external cusp is either absent or feebly developed. In the Gorilla, the last upper molar differs from the same tooth both in the Chimpanzee and in Man, in its greater size, being equal to the second except at the posterior part, which is slightly narrower, and in having both the posterior cusps, but especially the inner one, much more distinctly marked. The connecting ridge between the anterior internal and outer hind cusp is also present, but it is more transverse than the same ridge in the other molars. Each upper molar in both species of Troglodytes is implanted, as in Man, by one internal and two external fangs. These have not been observed connate in any of the Anthropoid Apes. In the lower jaw, the three molar teeth of the Gorilla are equal in size ; in the Chimpanzee they are nearly equal, the first being only slightly larger than the last. In both species, the four principal cusps, and especially the two inner ones, are more pointed and prolonged than in Man ; but the fifth posterior cusp is proportionably smaller. The fifth cusp is present in the second tooth ; it is usually absent in the second molar of Man. The crucial depression which separates the cusps is much less distinctly marked, and does not divide the ridge connecting the anterior pair, as in the human lower molar. Owing to the great development of the fifth posterior cusp, the crown of the last molar is of longer antero-posterior measurement. The implantation of these teeth resembles that in the human subject, except in the fact that the two roots of the second and third are never found connate in the Gorilla and Chimpanzee. The true molars in the Orangs (*Pithecus*) differ from the true molars of the Chimpanzees in their smaller relative size as compared with the premolars. Like the molar series in *Troglodytes*, they form a straight line in both jaws; and they have a similar implantation, by three roots in the upper and two in the lower jaw. In the superior series, the first and second teeth are furnished with four cusps; but they are less developed, and the depressions between them are shallower ; the connecting ridge between the antero-internal and the postero-external tubercles is either absent, or but very feebly developed, so that, as before stated, the sigmoid character of the elevated surface is not marked, as in Troglodytes ; the whole grinding surface is flatter, and minutely wrinkled. In the lower jaw, the first and second molars have three cusps along the outer side, and two on the inner. The last tooth of the series has two external cusps

the Lower-Jaw, and they are placed further back in the mouth, so that each is partly opposed to two of the Lower-Jaw. The second in the Upper-Jaw is smaller than the others, and the

distinctly marked, but the postero-internal tubercle is but feebly developed. The unworn surfaces of the lower molars are, like those of the upper, minutely wrinkled. Lastly, it would appear that, in the Orangs, it is not very unusual to find a fourth supernumerary molar, sometimes in the upper, sometimes in the lower jaw, on one or both sides. It is worthy of remark, that supernumerary teeth occurring in the human subject are generally incisors or canines.

The molar series in the true Carnivora presents a peculiar type. Its number varies in different genera, and the typical characters are most strongly marked in those which exhibit most strongly carnivorous propensities. The formula in the *Felidæ* is $p.\ \frac{3}{3}\ \vdots\ \frac{3}{3}$, $m.\ \frac{1}{1}\ \vdots\ \frac{1}{1}$. In the upper jaw, the first premolar (*p.* 2) is small and rudimentary; the second (*p.* 3) is much larger, implanted by two fangs, with a trenchant, conical crown, having its cutting edge slightly notched, indicating a division into three lobes, with a posterior basal ridge; the third (*p.* 4) is the upper "*dent carnassière*" of Cuvier—the scissor, or sectorial tooth. It is a large tooth, implanted by two fangs. The crown consists of a trenchant portion (the blade) which is divided into three lobes, of which the anterior is the smallest, the middle the longest and most pointed, and the posterior largest. On the inner side of the base of the anterior third of the crown, a thick tubercle is developed, which projects inwards. The two posterior thirds of the inner surface of the crown are smooth, and against this smooth portion works the outer surface of the blade of the lower carnassière. The fourth tooth (*m.* 1) is small and rudimentary. It is implanted behind and on the inner side of the carnassière. In the lower jaw, the premolar series is reduced to two by the absence of *p.* 2. Of the two premolars, the anterior is the smaller. They have both compressed, trenchant, conical crowns, divided by slight indentations in the cutting edge into three lobes, of which the middle is the largest. The second (*p.* 4) has a posterior basal ridge. The third tooth is the lower carnassière—the first molar. It is implanted by two fangs, and its crown is nearly equally divided into two trenchant lobes. It is to be remembered that the upper carnassière in the Carnivora always succeeds a deciduous tuberculate molar—it is, therefore, essentially a premolar; whilst the lower carnassière is developed behind the deciduous series—it is, therefore, the first true molar. The upper carnassière is always a little anterior to the lower. In the Hyænas, the molar formula is $p.\ \frac{4}{3}\ \vdots\ \frac{4}{3}$. $m.\ \frac{1}{1}\ \vdots\ \frac{1}{1}$. The teeth of the molar series in both jaws are larger and stronger than in the *Felidæ*, and an additional premolar tooth, *p.* 1 in the upper, *p.* 2 in the lower jaw, is retained. As in the *Felidæ*, the upper true molars are rudimentary. The premolars are remark-

first and second are placed directly under the Maxillary Sinus. I once saw the second Grinder naturally wanting on one side of the Lower-Jaw.

---

able for the strong basal ridges they present, which serve to protect the gums in the rough work of splitting and crushing bones. The first upper premolar is very small and conical, the second is much larger, and the third still larger, and of great strength. The strong cones of the second and third are each belted by a posterior and internal basal ridge, and the posterior part of the cone is also traversed by a longitudinal ridge. The fourth is the carnassière, and presents a blade divided into three lobes, and a strong internal trihedral tubercle. In the lower jaw, the first premolar (*p.* 2) fits into the interspace between the first and second premolars above. It presents a ridge anteriorly, and a broad basal talon behind. The second (*p.* 3) is the largest; it has an anterior and a posterior basal ridge, and its strong, rounded cone presents also an anterior and posterior vertical ridge. In the third (*p.* 4), the posterior basal ridge is developed into a small cone. The lower carnassière of the Hyæna resembles that of the *Felidæ*, except that the points of the two lobes are less produced, and from a small posterior talon a ridge is continued along the inner side of the base. In the family of the *Viverridæ* (Civets, Genets, Ichneumons, &c.), the molar series is $p.\ \frac{4}{4}.\ \frac{4}{4}$, $m.\ \frac{2}{2}.\ \frac{2}{2}$. This increase in the number of these teeth forms a link between the genus *Canis* and the genera already referred to. The sectorial teeth are more tubercular than trenchant. In some aquatic species, however, as in the sub-genus *Cynogale*, the premolars have compressed, triangular, trenchant, sharp-pointed crowns, the edges of which are minutely serrated, like the teeth of certain sharks. In the Indian Musangs (*Paradoxuri*), which are but little carnivorous in their habits, subsisting principally on the fruit of the palm-tree, these teeth manifest the opposite or tuberculate character. The lower sectorial (*m.* 1) has a pointed lobe developed on the inner side of its two anterior lobes; its posterior lobe is trituberculate, as in the dog. The crown of the last lower molar (*m.* 2) is oval, with four small tubercles, resembling *m.* 2 in the Dog. The molar formula in the genus *Canis* is $p.\ \frac{4}{4}.\ \frac{4}{4}$, $m.\ \frac{2}{3}.\ \frac{2}{3}$. The premolars have, strong, sub-compressed conical crowns; they increase in size from before backwards, the larger ones presenting one or two small posterior tubercles. The upper sectorial (*p.* 4) is of much larger size than the other teeth of the series. It is divided by a wide notch into two lobes, of which the anterior is much the larger and more produced; from the inner side of its base the tubercle is developed. The upper true molars are tuberculate; the second is very small. The lower sectorial (*m.* 1) has the blade formed by two conical lobes, of which the posterior is the larger; behind this the base of the crown extends into a broad quadrate trituberculate talon. The second molar presents two anterior opposite

The third Grinder is commonly called *Dens Sapientiæ*; it is a little shorter and smaller than the others, and inclined a little more inwards and forwards. Its body is nearly of the same

cusps, and a broad posterior talon. The third molar is very small. Professor Owen remarks—"The succession of two tubercular teeth behind the permanent sectorial tooth in the adult, or permanent dentition of the lower jaw, carries the genus *Canis* farther from the type of its order, and stamps it with its own proper omnivorous character, and this contributes to adapt the Dog to a greater variety of climates and food, and of other circumstances, all of which tend, in an important degree, to fit that animal for the performance of its valuable services to Man." In the Weasel tribe (*Mustelidæ*) there are usually three premolars in the upper jaw on each side; the Otter, however, has four. In the lower jaw on each side there are four or three premolars. The true molars are usually $\frac{1}{1}$: $\frac{1}{1}$:. In the Otter the sectorial and molar teeth present increased grinding surfaces, in relation to the coarser nature of their animal diet, and the necessity of crushing the bones of fish before they are swallowed. In the great Sea-Otter (*Enhydra*) the upper sectorial (*p.* 4) is remarkably modified; the two lobes of the blade are hemispheric tubercles. The last tooth (*m.* 1) is larger than the sectorial, and has a similar broad crushing form. The crown of the lower sectorial (*m.* 1) presents many gradations in this genus from the cutting form observed in the Weasel and Glutton, to the crushing type of the Ratel and Sea-Otter. In a South American Skunk, the second upper premolar is absent. The molar series, *p.* $\frac{3}{3}$: $\frac{3}{3}$:, *m.* $\frac{1}{2}$: $\frac{1}{2}$:, in the subfamily of Badgers (*Melidæ*) contrasts in its tuberculate and omnivorous character with the sectorial type found in the predaceous Weasel. The upper true molar in the European Badger attains an enormous size as compared with that of the same tooth in any of the preceding Carnivora; it presents three external tubercles, and an extensive horizontal surface traversed by a low ridge, and bounded by an internal belt. The molar series in the Bears (*Ursidæ*) is *p.* $\frac{4}{4}$: $\frac{4}{4}$:, *m.* $\frac{2}{3}$: $\frac{2}{3}$:; but the number of premolars presents some variation in different species. The true molars in both jaws present a tuberculate grinding surface. The teeth which correspond with the true molars in the Seals (*Phocidæ*) are more numerous than in the digitigrade family of Carnivores; in the upper jaw, they occasionally rise to the typical number of three on each side. The entire molar series in the Seals is usually five or six teeth on either side of the upper jaw, and five on each side of the lower. In some genera—as the Eared Seals (*Otariæ*), and Elephant Seals (*Cystophora*), they are each supported by a single fang, in other genera by two fangs, which are usually connate in the first or second teeth. The fangs of all the teeth in the Seals are of remarkable thickness. The crowns of the molars are generally compressed, conical, and pointed; they present the 'cingulum,' and

figure, but rather rounder, and its fangs are generally not so regular and distinct, for they often appear squeezed together; and sometimes there is only one fang, which makes the Tooth

anterior and posterior basal tubercles more or less developed. In a few, however, of the larger species the molars are simple and obtuse. This is especially the case in the Walrus, in which the molar series is reduced to a smaller number than in the true Seals, being, in the adult, three on each side in the upper jaw, and four in the lower. On turning to the extinct Carnivora, we find in *Machairodus* two premolars, p. 3 and p. 4, and one small rudimentary tubercular molar in the upper jaw; in the lower the molar series accords with that of *Felis*. The more ancient carnivorous forms of the older tertiary formations exhibit the typical formula of Placental Diphyodont dentition. In *Hyænodon* each of the three lower molars present the carnassial form as truly as the lower carnassière of the Felidæ. In another early carnivore (*Amphicyon*), with the same typical number of teeth, the molars have the tubercular conformation—the prototypes of the tubercular molars of the *Viverridæ* and *Canidæ*.

The study of the grinding teeth in the various vegetable feeders, is assisted and simplified by a reference to the forms of those teeth in the fossil remains of early herbivora, occurring in the eocene or most ancient tertiary formations. In the *Anoplotherium*, the crown of the upper molar is divided into a front and back lobe, by a valley which extends from its inner side two thirds across. A second valley, crosses its termination at right angles, and forms a crescentic depression in each lobe, concave towards the outer side of the crown—the side of the crown being impressed by two parallel excavations. There is a large conical tubercle at the entrance of the transverse valley on the inner side of the crown. This type is continued into the Dichodon, Dichobunes and existing Ruminants. The dentition of the Dichodon, an extinct genus, the remains of which have been found in the upper eocene of Hampshire, forms a transitional link between that of the Anoplotherium and the present Ruminants. The Dichodon presented the typical formula $p.\ \frac{4}{4}.\ \frac{1}{1}\ ,\ m.\ \frac{3}{3}.\ \frac{3}{3}.\ .$ The crowns of the first three premolars are extended from before backwards, each presents three progressively developed and pointed cusps on the same line; to the third in the upper jaw, a fourth inner and posterior cusp is added. The fourth premolar is thicker and shorter from before backwards, and has two pairs of cusps. The crowns of the upper true molars have also each two pairs of sharp-pointed cusps, and also a series of five low accessory points, developed from the outer part of the cingulum. The lower molars have crowns of the same complex character as those above, but the convex side of the principal cusps are turned in the opposite direction to those of the upper, and the accessory basal points are conical: it is much smaller than the rest of the Grinders. In the Upper-Jaw this Tooth has more variety than in the Lower, and is even smaller than the corresponding Tooth of the Lower,

developed from the inner instead of the outer side of the crown. Compared with the molar of the Anoplotherium, the outer lobes of that of Dichodon are thicker and sharper, the inner ones are more nearly equal to the outer, and are more distinctly divided from them; the transverse valley extends completely across the tooth, and is crossed by the antero-posterior doubly crescentic depression. In existing Ruminants, the cusps of the upper molars are less pointed and lower than in Dichodon. When worn, the crown presents the inner and outer pair of crescentic lobes of dentine, whilst the double crescentic valley separating them, containing a thicker layer of cement, forms two detached crescents. The premolar resembles one half of the true molar. The permanent formula of the grinding series in existing Ruminants is $p. \frac{3}{3}. \frac{3}{3}., m. \frac{3}{3}. \frac{3}{3}.$. The lower molar of the Ruminant resembles the upper one reversed. The single median longitudinal depression which in the upper tooth is internal, in the lower is external; whilst the two concavities of the outer side of the upper molar are repeated, although less deep on the inner side of the lower. The lower molars are thinner than the upper, and in the worn crown, the crescentic islands are narrower and less bowed. The outer contour of the grinding series in Ruminants is slightly zigzag, the anterior and outer angle of each tooth projecting beyond the posterior and outer angle of the tooth before it. The premolars form a continuous series with the molars. They are smaller and more simple.

In the Hyracotherium, another eocene herbivore, the grinding surface supports four principal cusps. Each transverse pair is connected by a ridge which supports a smaller conical tubercle, and the crown is girt with a cingulum. On this type are formed the upper grinders of the existing Hog tribe and of the Hippopotamus. The genus Sus is an instance of the retention of the typical formula $p. \frac{4}{4}. \frac{4}{4}., m. \frac{3}{3}. \frac{3}{3}.$. In the Wild Boar, the teeth of the grinding series increase in size from the first to the last. The first premolar has a simple, conical crown, thickest behind, and is implanted by two fangs. The second has a broader crown, with a posterior lobe, having a depression on its inner surface; each fang exhibits a tendency to division. The crown of the third is still broader, and it is implanted by four fangs. The fourth premolar has two principal tubercles, and some irregular vertical depressions on the inner half of the crown. The first true molar originally bears four principal cones, with smaller irregular sub-divisions, and an anterior and posterior ridge; but owing to its early development, the tubercles become worn down, and a smooth surface of dentine is exposed by the time the last molar is in place. A crucial depression, dividing four

and thence stands directly opposed to it; but for this circumstance the Grinders would reach farther back in the Upper-Jaw than in the Lower, which is not commonly the case.

---

cones, with more complex shallow divisions, and a larger tuberculate posterior ridge characterise the second true molar. The last molar is of great antero-posterior extent, owing to the development of the posterior ridge into a cluster of tubercles. The four primary cones are distinguishable on the fore part of the surface of the crown. In the Warthogs (*Phacochœrus*), the molar series is reduced by the suppression of $p. 1$ and $p. 2$. The last true molar in the Wart-hogs is very remarkable for its extent and complexity. It is, perhaps, the most peculiar and complex tooth in the Mammalian series. The surface of the crown presents three series of enamel islands, ranged in the long axis of the grinding surface; each row consists of eight or nine islands. Those of the middle row are elliptic and simple; those of the inner and outer rows are sometimes sub-divided into smaller islands. Each island or lobe consists of an abraded column of dentine encased by thick enamel, and the whole are united together to form the crown by abundant cement which fills up the interspaces, and gives an external covering to the whole tooth. The molar series in the Hippopotamus consists of $p. \frac{4}{4}\frac{4}{4}, m. \frac{3}{3}\frac{3}{3}$. The first premolar has a simple conical crown and a single root. It comes early into place at some distance in advance of the second, and is soon shed. In the existing Hippopotamus, the other premolars and molars form a continuous series; but in the *Hippopotamus Major*, of the pliocene strata, the second premolar was in advance of the third by an interval equalling its own breadth. The third and fourth premolars retain the conical form, but are of larger size, and present one or two longitudinal grooves on the outer surface, which give the crown, when worn, a lobed appearance. The true molars are each divided into two cones or lobes by a wide transverse valley. Each cone is again sub-divided by a narrow, antero-posterior cleft, into two half cones, with their flat sides next each other. When moderately worn, each half cone presents a trefoil of enamel; but when worn to the base, the surface of each lobe presents a quadrilobate figure. The crown of the third molar has a fifth smaller cone developed behind the two normal pairs of half cones.

In the Palæotherium, another eocene herbivore, the crown of the molar is divided into an anterior and posterior part, by an oblique fissure, continued from near the middle of the inner surface of the crown obliquely outwards and forwards, two thirds across the tooth. This fissure enlarges at its termination, so as to mark a division of the anterior part of the crown into internal and external lobes. The posterior division of the crown is likewise divided into inner and outer lobes by a short, wide valley or fissure, which extends forward from the

In the Upper-Jaw this third Grinder is turned but a very little outwards; is frequently inclined somewhat backwards; and it projects over that of the Under-Jaw. It oftener becomes loose than any of the other Teeth.

---

posterior surface. This type is the fundamental pattern of the upper molars of the Horse and Rhinoceros. The formula of the molar series in the Horse is $p. \frac{3}{3}. \frac{3}{3}. m. \frac{3}{3}. \frac{3}{3}.$. The crowns of the upper molars bear some resemblance to those of the Ruminants in their complexity. The resemblance of the crowns of the lower molars to those of the Rhinoceros and Palæotherium is more marked. The grinders of the Horse may be distinguished from the complex teeth of other *Herbivora* of an equal size by the great length of the tooth before it divides into fangs. Abrasion of the crown goes on to a considerable extent before the division begins; hence, except in old horses, a considerable portion of the whole molar is implanted by an undivided base. In the Rhinoceros, the likeness of the crown of the molar to that of the Palæotherium is very obvious. The differences in the upper molar are chiefly that two concavities which exist on the outer side of the crown in Palæotherium are almost levelled in the Rhinoceros; that the termination of the oblique transverse fissure in the Rhinoceros is more expanded, and in some species bifurcates and deepens, so that in the worn crown one branch may form an insulated circle of enamel. The posterior valley is also deeper and more extended. A basal ridge girts the internal and anterior and posterior sides of the crown. The formula of the grinding series in the Rhinoceros is $p. \frac{4}{4}. \frac{3}{3}. m. \frac{3}{3}. \frac{3}{3}.$.

A fourth type is furnished by the molars of the Lophiodon. They are nearly allied, in configuration, to those of Palæotherium, but they have a more decided transversely ridged character. This configuration obtains in the Dinotherium and the existing Tapirs.

The Proboscidean Family. Certain extinct species of proboscidians received from Cuvier the name of "Mastodon," in consequence of the peculiar conformation of their grinding teeth, which appeared to place them in a distinct group from the existing Elephant. Subsequently, however, other fossil remains of proboscidians have been discovered in the tertiary deposits of tropical Asia, which establish the transitional characters between the lamello-tuberculate teeth of the elephants and the mammilated molars of the typical Mastodons, showing that the distinctions deducible from their molar teeth rather establish differences of species than of genera. In the *Mastodon giganteus* and *M. angustidens*, the grinding surface of the molar, instead of being cleft into numerous thin plates, as in the Elephant, was divided into wedge-shaped, transverse ridges, the summits of which were again divided into conical protuberances, more or less resembling the teats of a cow. The crown of the

They are placed under the posterior part of the Maxillary Sinus, and there the parts which compose the Sinus are thicker than in the middle. The variations as to the natural number of the Teeth, depend commonly upon these *Dentes Sapientiæ*. (*p*)

---

tooth is formed of dentine, thickly coated with dense and brittle enamel. A thin covering of cement is continued from the fang upon the crown, but it does not fill up the interspaces of the divisions of the crown as in the Elephant. In the Mastodons, there were three deciduous molars (*d. m.* 2, 3, and 4) on each side, in both jaws ; the second was replaced by one premolar (*p.* 3), and there were three true molars on each side, both above and below. The grinders of the Proboscidians follow each other, from before backwards, at long intervals, and are never simultaneously in place. In the Mastodons not more than three were in use at any period on one side of either jaw ; all the molars, save the penultimate, were shed by the time the last molar had cut the gum, and in the old Mastodon, the dentition was at last reduced to *m.* 3 on each side of both jaws. In the existing Elephants, the grinders are $d.\ m.\ \frac{3}{3}$ $m.\ \frac{3}{3}$; no premolar replaces either of the deciduous molars. The grinding teeth are large and complex, and there is never more than one, or two partially, in place and use at the same time. The series is constantly in progress of growth and destruction, shedding and replacement. No premolar succeeding the deciduous molars, all the grinding teeth follow each other horizontally from behind forwards. The structure of the Elephants tooth has been before described (*Vide* note 1., p. 192, vol. iii.) In the extinct species which, at one time, was a denizen of Northern Europe and Asia—the Mammoth—(*Elephas primigenius*), the enamelled plates were more numerous in proportion to the size of the crown, and thinner than in the existing Asiatic species. In the African Elephant, on the other hand, the plates of enamel-covered dentine are fewer and thicker : they expand from the margins to the centre of the tooth, and present a lozenge form when worn down by mastication. The final blending of the plates by a common base of dentine does not take place simultaneously along the whole extent of the tooth in the Indian Elephant ; the anterior plates which are first formed, are worn down, and the base of dentine is exposed, whilst the posterior divisions of the crown are still distinct, adhering only by cement. The African Elephant, by the complete basal confluence of the plates before the anterior ones are worn out, manifests a closer affinity to the Mastodon. (1)

(*p*) [In the Melanian varieties of Man, the molar teeth are of large size, and the fangs of the wisdom and penultimate molars are not, as a rule, connate or conjoined. The great size of the molar teeth is most

---

(1) Professor Owen, Article Odontology, Enyclopædia Britannica, 8th edit.

Thus from the *Incisores* to the first Grinder, the Teeth become gradually thicker at the extremity of their bodies, and smaller from the first Grinder to the *Dens Sapientiæ*. From the *Cuspidatus* to the *Dens Sapientiæ* the fangs become much shorter; the *Incisores* are nearly of the same length with the *Bicuspides*. From the first *Incisor* to the last Grinder, the Teeth stand less out from the sockets and Gums.

The bodies of the Teeth in the Lower-Jaw are turned a little outwards at the anterior part of the Jaw, and thence, to the third Grinder, they are inclined gradually more inwards. The Teeth in the Upper-Jaw project over those of the Under, especially at the fore-part, which is owing to the greater obliquity of the Teeth in the Upper-Jaw; for the circle of the sockets is nearly the same in both Jaws. This oblique situation, however, becomes gradually less, from the *Incisores* backwards, to the last Grinder, which makes them gradually project less in the same proportion.

---

remarkable in the Australian variety, in which race also the wisdom teeth attain large relative dimensions, and are generally distinguishable from those of Europeans, not only by a complex implantation by distinct fangs, but by the fuller development of their posterior tubercles. A careful examination, however, of the dentition in a large collection of crania will prove that the dimensions of the molar teeth in Australian skulls vary; that in some cases they are equalled in size by the same teeth in other races, and that in exceptional instances the development and implantation of the upper third molar does not differ from the ordinary standard. In the Museum of the Royal College of Surgeons is an Australian skull, in which the upper wisdom exactly resembled the same tooth in the cranium of a Celtic Scot, with which it was compared. It was considerably smaller than the penultimate, whilst the diminution, as in Europeans, principally depended on the minor development of the posterior internal tubercle. In another instance, in the same race, the writer found that the three fangs of the upper third molar were conjoined, and he has noticed the same thing in a skull of an allied variety, the Papuan of New Guinea. In the West Coast African Negro, the molars also usually attain large, but not exceptional, dimensions. In skulls of the Hottentot and Bushman race which have come under the writer's notice, the development of these teeth has not been above the ordinary standard. (1)]

---

(1) 'On the Teeth in the Varieties of Man,' DENTAL REVIEW, 1860.

The Teeth in the Upper-Jaw are placed farther back in the circle than the corresponding Teeth of the Lower; this is owing to the two first *Incisores* above being broader than the corresponding *Incisores* below. All the Teeth have only one fang, except the Grinders, each of which has two in the Lower-Jaw, and three in the Upper.*

The fangs bear a proportion to the bodies of the Teeth; and the reason is evident, for otherwise they would have been easily broken, or pushed out of their sockets. The force commonly applied to them is oblique, not perpendicular; and they are not so firmly fixed in the Upper-Jaw, that is, the Alveolar Process in that is not so strong as in the Under-Jaw: it is perhaps on this account, that the Grinders in that Jaw have three fangs.

This particular structure in the Alveolar Process of the Upper-Jaw, is perhaps to give more room for the *Antrum Highmorianum;* on this supposition the fangs must be made accordingly, *i.e.* so that they shall not be pushed into that cavity; now, by their diverging, they inclose as it were the bottom of the *Antrum*, and do not push against its middle, which is the weakest part; and the points of three diverging fangs will make a greater resistance (or not be so easily pushed in) than if they were placed parallel. If there had been only two, as in the Lower-Jaw, they must have been placed opposite to the thinnest part of the *Antrum ;* and three points placed in any direction but a diverging one, would have had here much the same effect as two; and as the force applied is endeavouring to depress the Tooth, and push it inwards, the innermost fang diverges most and is supported by the inner wall of the *Antrum*. That all

---

* Those Anatomists who allow the Teeth to have more fangs, have been led into a mistake, I suppose, by often observing two canals in one fang; and thence concluded, that such a fang was originally two, and that these were now growing together. (*q*)

(*q*) [This observation of Hunter's is itself a mistake. The fang of the upper premolar really consists of two connate fangs.]

this weakness in the Upper-Jaw is for the increase of the *Antrum* is probable, because all the Teeth in the Upper-Jaw are a good deal similar to those in the Lower, excepting those that are opposite to the Maxillary Simus; and here they differ principally in the fangs, without any other apparent reason; and what confirms this, is, that the *Dentes Sapientiæ* in both Jaws are more alike than the other Grinders; for this reason, as I apprehend, because the *Dens Sapientiæ* in the Upper-Jaw, does not interfere so much with the Maxillary Sinus.

What makes it still more probable that the two first superior Grinders have three fangs on account of the Maxillary Sinus, is, that the two Grinders on each side of the Upper-Jaw, in the child, have three fangs, and we find them underneath the *Antrum;* but those that succeed them have only one fang, as in the Lower-Jaw; but by that time the *Antrum* has passed further back, or rather the arch of the Jaw has projected, or shot forwards, as it were, from under the *Antra*, so that the Alveolar Processes that were under the *Antrum* at one age, are got before it in another.

That the edge of every fang is turned towards the circumference of the Jaw, in order to counteract the acting power, we shall see when we consider the Motion of the Jaw, and the Use of the Teeth.

OF THE ARTICULATION OF THE TEETH.

The fangs of the Teeth are fixed in the Gum and Alveolar Processes, by that species of Articulation called Gomphosis, which, in some measure, resembles a nail driven into a piece of wood.

They are not, however, firmly united with the Processes, for every Tooth has some degree of motion; and in heads which have been boiled or macerated in water, so as to destroy the Periosteum and adhesion of the Teeth, we find the Teeth so loosely connected with their sockets, that all of them are ready to drop out, except the Grinders, which remain as it were hooked from the number and shape of their fangs.

OF THE GUMS.

The Alveolar Processes are covered by a red vascular substance, called the Gums, which has as many perforations as there are Teeth; and the neck of a Tooth is covered by, and fixed to this Gum. Thence there are fleshy partitions between the Teeth, passing between the external and internal Gum, and, as it were, uniting them; these partitions are higher than the other parts of the Gum, and thence form an arch between every two adjacent Teeth. The thickness of that part of that Gum which projects beyond the sockets is considerable; so that when the Gum is corroded by disease, by boiling, or otherwise, the Teeth appear longer, or less sunk into the Jaw. The Gum adheres very firmly in a healthful state both to the Alveolar Process and to the Teeth, but its extreme border is naturally loose all around the Teeth. The Gum, in substance, has something of a cartilaginous hardness and elasticity, and is very vascular, but seems not to have any great degree of sensibility; for though we often wound it in eating, and in picking our Teeth, yet we do not feel much pain upon these occasions; and both in infants and old people, where there are no Teeth, the Gums bear a very considerable pressure, without pain. (*r*)

The advantage arising from this degree of insensibility in the Gums is obvious, for till the child cuts its Teeth, the Gums are to do the business of Teeth, and are therefore formed for this purpose, having a hard ridge running through their whole

---

(*r*) [The gum (gingiva) owes its resistance principally to the hardness of the subjacent parts; in itself it is soft. It is composed of a firm, dense, whitish, submucous tissue, and a proper mucous membrane, covered by a layer of pavement epithelium. The mucous membrane bears papillæ of considerable size (0·15''' to 0·3'''). The epithelium is 0·23''' to 0·4''' in thickness between the papillæ. According to Kölliker, there are no glands in the gum. He observes: "We must be careful not to take for the orifices of glands, certain rounded depressions of the epithelium, 0·08''' to 0·15''' in diameter, with cornified epithelial cells."(1)]

---

(1) Kölliker, *op cit.*, p. 300.

length. Old people, who have lost their Teeth, have not this ridge. When in a sound state, the Gums are not easily irritated by being wounded, and therefore are not so liable to inflammation as other parts, and soon heal.

The Teeth being united to the Jaw by the Periosteum and Gum, have some degree of an yielding motion in the living body. This circumstance renders them more secure; it breaks the jar of bony contact, and prevents fractures both of the sockets and of the Teeth themselves.

OF THE ACTION OF THE TEETH ARISING FROM THE MOTION OF THE LOWER-JAW.

The Lower-Jaw may be said to be the only one that has any motion in mastication; for the Upper-Jaw can only move with the other parts of the head. That the Upper-Jaw and head should be raised in the common act of opening the mouth, or chewing, would seem, at first sight, improbable; and from an attentive view of the mechanism of the joints and muscles of those parts, from experiment and observation, we find that they do not sensibly move. We shall only mention one experiment in proof of this, which seems conclusive: let a man place himself near some fixed point, and look over it, to another distant and immoveable object, when he is eating. If his head should rise in the least degree, he would see more of the distant object over the nearest fixed point, which in fact he does not. The nearer the fixed point is, and the more distant the object, the experiment will be more accurate and convincing. The result of the experiment will be the same, if the nearest point has the same motion with the head; as, when he looks from under the edge of a hat, or any thing else put upon his head, at some distant fixed object. We may conclude then that the motion is entirely in the Lower-Jaw: and, as we have already described both the articulation and the motion of the bone, we shall now explain the action of mastication, and, at the same time, consider the use of each class of Teeth.

With regard to the action of the Teeth of both Jaws, in

mastication, we may observe, once for all, that their action and re-action must be always equal, and that the Teeth of the Upper and Lower-Jaws are complete, and equal antagonists both in cutting and grinding.

When the Lower-Jaw is depressed, the Condyles slide forwards on the eminences; and they return back again into the cavities, when the Jaw is completely raised.

This simple action produces a grinding motion of the Lower-Jaw, backwards on the Upper, and is used when we divide any thing with our fore Teeth, or *Incisores*. For this purpose, the *Incisores* are well formed; as they are higher than the others, their edges must come in contact sooner; and as the Upper project over the Under, we find in dividing any substance with them, that we first bring them opposite to one another, and as they pass through the part to be divided, the Lower-Jaw is brought back, while the Incisors of that Jaw slide up behind those of the Upper-Jaw, and of course pass by one another. In this way they complete the division, like a pair of scissars; and at the same time they sharpen one another. There are exceptions to this; for these Teeth in some people meet equally, viz. in those people whose Fore-Teeth do not project further from the Gum, or socket, than the back Teeth; and such Teeth are not so fit for dividing; and in some people the Teeth of the Lower-Jaw are so placed, as to come before those of the Upper-Jaw; this situation is as favourable for cutting as when the over-lapping of the Teeth is the reverse, except for this circumstance, that the Lower-Jaw must be longer, and therefore its action weaker.

The other motion of the Lower-Jaw, viz. when the lateral Teeth are used, is somewhat different from the former. In opening the mouth, one Condyle slides a little forwards, and the other slides a little further back into its cavity; this throws the Jaw a little to that side, just enough to bring the lower Teeth directly under their corresponding Teeth in the Upper-Jaw: this is done, either in dividing, or holding of substances; and these are the Teeth that are generally used in the last mentioned action. When the true grinding motion is to be performed, a greater degree of this last motion takes place; that is, the Condyle of the opposite side is brought farther forwards, and the Condyle of the same side is drawn farther back into the cavity of the Temporal Bone, and the Jaw is a little depressed. This is only preparatory for the effect to be produced; for the moving back of the first mentioned Condyle into the socket, is what produces the effect in mastication.

The lateral Teeth in both Jaws are adapted to this oblique motion; in the Lower they are turned a little inwards, that they may act more in the direction of their axis; and here the Alveolar Process is strongest upon the outside, being there supported by the ridge at the root of the Coronoid Process. In the Upper-Jaw the obliquity of the Teeth is the reverse, that is, they are turned outwards, for the same reason; and the longest fang of the Grinders is upon the inside, where the socket is strengthened by the bony partition between the *antrum* and nose. Hence it is, that the Teeth of the Lower-Jaw have their outer edges worn down first; and, *vice versa*, in the Upper-Jaw.

GENERAL COMPARISONS BETWEEN THE MOTION OF THE JAW IN YOUNG AND IN OLD PEOPLE.

In children who have not yet Teeth, there does not seem to be a sliding motion in the Lower-Jaw. The articular eminence of the Temporal Bone is not yet formed, and the cavity is not larger than the Condyle; therefore the centre of motion in such, must be in the Condyle. In old people, who have lost their Teeth, the centre of motion appears to be in the Condyles, and the motion of their Jaw to be only depression and elevation. They never depress the Jaw sufficiently to bring the Condyle forwards on the eminence, because in them the mouth is sufficiently opened when the Jaw is in its natural position.

Hence it is that in old people, the gums of the two Jaws do not meet in the fore-part of the mouth, and they cannot bite at that part so well, as at the side of the Jaw; and, instead of the grinding motion, which would be useless, where there are no Grinders; they bruise their food rather by a simple motion of the Jaw upwards and downwards.

It is from the want of Teeth in both those ages, that the face is shorter in proportion to its breadth. In an old person, after the Teeth are gone, the face is shorter, while the mouth is shut, by almost the whole lengths of the Teeth in both Jaws; that is, about an inch and an half.

From the want of Teeth too, at both those ages, the cavity of the mouth is then smaller; and the tongue seems too large and unmanageable, more especially in old people. In these last we observed also, that the chin projects forwards, in proportion as the mouth is shut; because the basis of the Lower-Jaw (which is all that now remains) describes a wider circle than the Alveolar Process in younger people. The Jaws do not project so much forwards in a child, as in an adult; hence the face is flatter, especially at the lower part. In proportion as the last Grinders are produced, the sides of the curve formed by the Jaws become longer, and push forward the fore-part, none of the additional part passing backwards. The fore-part also continues nearly of the same size, so that the whole Jaw is longer in proportion to its breadth, and projects farther forwards.

OF THE FORMATION OF THE ALVEOLAR PROCESS.

Having considered the Alveolar Processes in their adult, or perfect state, we shall next examine and trace them from their beginning.

We observe the beginning of the Alveolar Process at a very early period. In a fœtus of three or four months it is only a longitudinal groove, deeper and narrower forwards, and becoming gradually more shallow and wider backwards: instead of bony partitions, dividing that groove into a number of sockets, there are only slight ridges across the bottom and sides, with intermediate depressions, which mark the future *Alveoli*.

In the Lower-Jaw the vessels and nerves run along the bottom of this Alveolar cavity, in a slight groove, which afterwards becomes a complete and distinct bony canal.

The Alveolar Process grows with the Teeth, and for some time keeps the start of them. The ridges which are to make the partitions shoot from the sides across the canal, at the mouth of the cell, forming hollow arches: this change happens first at the anterior parts of the Jaws. (*s*) As each cell becomes

---

(*s*) [A very full description of the formation of the alveoli is given by Drs. Robin and Magitot, in their "Treatise on the Genesis and Development of the Dental Follicles." According to these observers, the first traces of ossification in the lower jaw appear near the middle of the inferior maxillary cartilage; they are elongated in form, and show themselves about the thirty-fifth day of intra-uterine existence. The ossifying process rapidly invades the whole cartilage. Two very thin ridges then appear on the upper margin of the bone; these ridges do not exist in the cartilaginous state, but are formed according to the mode of ossification called "invasion." At an early period, they give the bone the appearance of being formed by two parallel osseous bands. These ridges limit the groove in which the dental follicles originate. The groove extends from the anterior edge of the ascending branch of the maxillary, encroaching a little on its inner face, as far as the anterior extremity of the corresponding branch of the maxillary; consequently, the whole of its contents can be removed in one piece. The groove is of considerable depth. In Man, the solid part of the bone, from the canine as far as the symphisis, becomes higher than the groove is deep about the beginning of the fourth month. At the level of the molars, and in relation to the axis of the lower maxillary, the groove is situated inside the latter, but passes round it, in order to be continued on the side of the outer face in the whole portion containing the follicles of the canine and the incisors. The groove is widened, as if swollen in blisters about its posterior third, narrow in front, and more suddenly contracted behind. It opens at the inner face of the ascending branch of the maxillary by a fissure-shaped opening, broadened and rounded at the level of the bottom of the groove, and narrow above, where it soon closes; there then remains only the lower part of the slit which forms the posterior dental foramen, or posterior orifice of the dental canal, which is traversed by the vessels and nerves of the same name; they make part of the contents of the groove. The dental vessels and nerves are contained in a slight furrow at the bottom of the groove. The furrow is smooth and regular, and, at a later period, becomes the dental canal. From the time of the genesis of the dental follicles, the inner surface of each lamina forming the sides of the groove becomes thickened

deeper, its mouth also grows narrower, and at length is almost but not quite, closed over the contained Tooth.

The disposition for contracting the mouth of the cell, is

---

at intervals by little vertical projections, which stand facing each other on each side. These projections increase, and ultimately join each other, forming complete partitions, which divide the groove into cells, or alveoli. In Man, this takes place at an advanced period of gestation. As late as the ninth month of fœtal life, the contents of the groove can be raised in one piece. The partitions bridge over the lower part of the groove containing the dental vessels and nerves. The rudiments of the partitions always unite, in the first place, between the first molar and the canine, and then between the latter and the second incisor. The rudiments of the first partition join to form a narrow, slender, and thin lamina, passing over the nerves and vessels at the bottom of the groove towards the end of the fourth month. The prolongations between the canine and second incisor become united near the sixth month. They unite between the two incisors near the seventh month, but not between the molars until after birth. Before the projections unite throughout their whole height, they form a very slender bridge immediately above the vessels. This bridge grows in height by prolongations from the walls of the groove. The partitions are not completed until after birth. The following is the account given by the same observers of the formation of the alveoli in the upper jaw. About the fifty-fifth day (in the human subject) there appear at the outer edge of the superior maxillary bone two thin crests, an external and an internal one; these crests limit a groove, which has at first the appearance of a shallow furrow. Shortly after, a similar groove is formed in the same manner upon the inter-maxillary. In the posterior, two-thirds of the groove in the superior maxillary, the suborbital vessels, and nerves are lodged. The external lip of the dental groove passes above them, near its anterior extremity, in order to allow them to reach under the skin. "This groove is thus produced immediately below the eye, a place occupied at that time by the edge of the upper maxillary, and which the nerves and vessels continue to occupy. The same thing occurs in the lower maxillary, where the groove appears before that of the other jaw. The groove of the upper maxillary is common to the dental follicles, which are about to appear, and to the vessels which remain suborbital." The bottom of the groove becomes in the course of development the suborbital canal, just as in the lower maxillary it becomes the dental canal, but the separation between the canal and the groove takes place much sooner in the superior than in the inferior maxilla. "Nevertheless, it is at the bottom of this groove, against the suborbital nerves and vessels, and consequently in the part which afterwards becomes the suborbital canal, that the dental follicles also

chiefly in the outer plate of the bone, which occasions the contracted orifices of the cells to be nearer the inner edge of the Jaw. The reason, perhaps, why the bone shoots over, and almost covers the Tooth, is that the Gum may be firmly supported before the Teeth have come through. (*t*)

---

originate. Those of the molars are, however, the only ones that originate here, for the canal belongs only to the maxillary, and not to the incisive bone, and because it is already separated from the groove in front when the follicle of the canine makes its appearance." From the third month, the posterior half of the groove, which was confounded with the suborbital canal, becomes closed, and separates the dental follicles placed at the level of the vessels and nerves." The external and internal plates which bound the groove are thin, fragile, and slightly undulated. The plates sink in a little opposite the intervals between the dental follicles. About the beginning of the fourth month, the rudiments of partitions begin to form as in the lower jaw, but they are produced from the bottom as well as the sides of the groove. "After the seventh month, the partitions reach in height nearly but not quite the level of the edges of the groove between the first molar and the canine, as well as between the incisors; the rest of the groove remains undivided and common to the two molars. At this period, the nerve and suborbital vessels, which start at first at the level of the line of contact between the canine and the second incisor, show their orifice of egress at the level of the rudimental partition interposed between the first molar and the canine. That part of the upper maxillary, which separates the canal of the bottom from the follicles is, at this time, only in the condition of a thin osseous plate, not thicker at most than from a quarter to half a millimetre, and perforated by one or two very small orifices intended for the passage of the vessels destined for these organs." (1)]

(*t*) [Mr. Tomes, who has given a very minute and accurate description of the changes which take place in the child's jaw from the time of birth to the fifth year, states that the partial closure of the alveoli commences between the second and third month after birth. He says that, when the third month has been attained, " the alveoli exhibits a considerable change in character, their depth has increased, and the free edges which were before open, so that in a macerated preparation the teeth readily fall out, are now turned inwards towards the median line of the alveolar ridge, thereby contracting the orifices, and affording protection to the

---

(1) Robin and Magitot in the 'Journal de la Physiologie de l'Homme et des Animaux.' Edited by Dr. E. Brown-Sequard, 1860. Translated in the 'Dental Cosmos.'

The *Alveoli* which belong to the adult Grinders, are formed in another manner; in the Lower-Jaw they would seem to be the remains of the root of the Coronoid Process; for the cells are formed for those Teeth in the root of that Process; and in proportion as the body of the bone, and the cells already formed, push forwards from under that Process, the succeeding cells and their Teeth are formed, and pushed forward in the same manner.

In the Upper-Jaw there are cells formed in the tubercles for the young Grinders, which at first are very shallow, and become deeper and deeper as the Teeth grow; and they grow somewhat faster, so as almost to inclose the whole Tooth before it is ready to push its way through that inclosure and Gum. There is a succession of these, till the whole three Grinders are formed.(*u*)

---

enclosed teeth, which are no longer liable to fall out when the bone is examined." At the age of six months he has found the inversion of the edges of the alveoli, and consequent narrowing of the apertures less pronounced. He adds, "the increased size of the alveolar orifices must be regarded as the first of those changes which precede the eruption of the teeth." (1)]

(*u*) [At the time of birth, a large open socket exists, posterior to the alveolus for the first temporary molar in both jaws. In the upper jaw, this, at that time, has a very imperfect posterior wall. The rudiments of a septum, however, intended to divide the large socket into two alveoli, one for the second temporary, and the other for the first permanent molar, are present. The division of the socket usually takes place a little earlier in the lower than in the upper jaw. At the age of eight months, the septum between the second milk and first permanent molar is still imperfect. In the upper jaw the posterior wall of the crypt of the first permanent molar is in progress of development, but not complete. At twelve months, the alveolus of the first permanent molar in the upper jaw has become more perfect. The alveoli of these teeth, in both jaws, communicate with the surface by small openings, in a line with the alveoli of the other teeth. At eighteen months, "the first permanent molars lie deep within their respective sockets, the orifices of which, in the lower jaw, are contracted by the inversion of the outer alveolar plate, and the base of the coronoid process, the teeth at this time being placed with their posterior two-thirds, internal to that portion of the jaw. The posterior edge of the socket is brought forward

---

(1) Tomes's 'System of Dental Surgery,' pp. 13—16.

OF THE FORMATION OF THE TEETH IN THE FŒTUS.

The depression, or first rudiments of the *Alveoli* observable in a Fœtus of three or four months, is filled with four or five little pulpy substances, which are not very distinct at this age.(*r*)

---

over the back part of the crown, to the extent of one-fourth of its antero-posterior dimensions. On the upper surface of this, within a line of its edge, a depression in the bone may be seen. This is a commencement of a crypt for the second permanent molar. The corresponding teeth of the superior maxillæ occupy the tuberosity, the posterior part of which is extremely thin, and in the median line imperfect. This gives a long and curved opening to the socket, and a posterior direction to its further half. In the upper jaw we have, as yet, no indications in the bone, of preparations for the lodgement of the second permanent molar." (1) At the age of forty months, the opening of the crypt of the first upper permanent molar has become smaller, and is directed downwards instead of downwards and backwards. Behind it on the posterior surface of the tuberosity is a depression marking the commencement of the crypt for the second permanent molar. Nine months later, the crypts for the second permanent molars have become larger, with well-defined margins. "In the upper jaw they look backwards towards the pterygoid plates of the sphenoid; in the lower, upwards and a little inwards, their floors lying immediately over the inferior dental canal, near its commencement." About the age of seven years, when the first permanent molars have gained the level of the temporary teeth, the second upper molars, which at first were directed backwards, now begin to descend into the dental line, and are directed obliquely downwards and backwards. Between the age of twelve and thirteen, the second permanent molars are advancing towards the surface of the gums, and the crypt for the third molars hold the positions which those for the second occupied when the first molars were emerging from the alveoli. (2)]

(*v*) [The dental pulps or bulbs. Two opinions are current with respect to the mode of origin of the dental pulps which may be here briefly stated. One is that the dental pulp commences as a papilla on the free surface of the mucous membrane covering the edges of the maxillary arches. This is held by Arnold and Goodsir, who came to their conclusions independently of each other. The other is that the dental pulp, together with the follicle in which it is afterwards contained, and the "enamel organ," have their origin in the submucous tissue which fills the groove in the maxillary. This opinion is sustained by the researches of Robin and Magitot.

---

(1) Tomes, Op. cit., p. 30.
(2) Tomes, Op. cit., pp. 5—103.

About the fifth month both the processes themselves and the

The following description of the formation of the different parts of the dental follicle has been generally adopted on the authority of Goodsir. About the sixth week of fœtal life a groove, *the primitive dental groove*, is formed along the edge of the jaw, in the mucous membrane of the gum; from its floor ten papillæ rise in succession in each jaw. The papillæ appear in the following order:—the first milk molar (7th week); the canine (8th week); the two incisors (about the 9th week); the second milk molar (10th week). The papillæ in the upper jaw appear a little before those in the lower. With the appearance of the last papilla the *papillary stage* terminates. The next stage is the *follicular*. This consists in the thickening and deepening of the margins of the dental groove, and the separation of the papillæ by the formation of membranous partitions which pass across from one margin of the groove to the other. This stage is completed about the fourteenth week. In the early part of this stage the papillæ grow rapidly and begin to assume peculiarities of form according with those of the crowns of the future teeth. The follicular stage is converted into the *saccular* about the end of the fifteenth week. The change consists in the formation of small laminæ or opercula of membrane which project from the sides of the follicle, and at last, by meeting and cohering, convert the follicle into a closed sac. The lips of the dental groove also become thickened and unite, so that the groove is at length entirely obliterated. The number of opercula in each follicle is said to vary with the character of the future tooth, there being four or five in the case of the molar; two, one external and one internal, in that of the incisor; and three, two internal and one external for the canine. About the fourteenth week a lunated depression is formed behind each milk follicle; these escape the general adhesion of the margins of the groove. In them the pulps of the ten anterior permanent teeth are subsequently developed; they are called by Mr. Goodsir "cavities of reserve." The formation of the pulp and the closure of the cavity by opercula is similar to that already described in the case of the follicles of the milk teeth. The pulp of the first permanent molar arises as a papilla about the sixteenth week in a portion of the primitive dental groove which remains unclosed behind the sac of the second milk molar. The papilla seated at the bottom of this "*posterior cavity of reserve*" becomes shut off and enclosed in a sac by the formation and cohesion of opercula, whilst the upper portion of the cavity lengthens backwards, and forms a second cavity of reserve in which the papilla for the second molar tooth appears about the seventh month after birth. Ultimately the same change again occurs, and gives rise to the formation of the papilla and sac of the wisdom tooth, which takes place about the sixth year.

Considerable doubt, however, has been thrown by various observers on the account given by Arnold and Goodsir. Valentine and Rasch-

pulpy substances become more distinct; the anterior of which

---

kow (1) before the publication of Goodsir's observations had contradicted those of Arnold; and more recently M. Natalis Guillot has given an entirely different account of the formation of the tooth follicles, in which he denies their origin on the surface of the mucous membrane, neither does he admit that they depend on any folding of the mucous membrane which has carried the germ from the surface into the cavity of the alveolus.(2) More recently the researches of Robin and Magitot have led them to a similar conclusion. According to these observers the dental follicles and bulbs originate in the submucous tissue which extends to the bottom of the osseous groove in the maxillaries. The term follicle is made to include the bulb, the organ of the enamel and the follicular wall; of these parts the bulb originates first, then the follicular wall, which subsequently closes over, and, lastly, the organ of the enamel as soon as the follicle is closed. The follicles of the lower jaw precede those of the upper. In the human foetus the appearance of the follicles is nearly in the same order as the eruption of the corresponding teeth in each jaw. The internal incisor and the anterior molar first. These are followed closely by the external incisor, then the posterior molar, and, lastly, the canine. The changes which take place in the dental follicles follow an order in accordance with that of the appearance of the follicles,—*i.e.*, in the follicle which appears first the follicular wall closes first, the organ of the enamel appears first, the ivory and enamel originate first. The first follicle in the lower jaw appears about the sixtieth day, that in the upper about the sixty-fifth day. The number of follicles of the first dentition in the lower jaw is complete about the seventy-fifth day; in the upper jaw about five days later. The follicles of the first permanent molars appear in the lower jaw about the eighty-fifth day; in the upper from the ninetieth to the ninety-fifth. The follicles of the teeth of replacement appear, some a little before birth or sometimes later, the others at longer or shorter intervals after birth. The bulbs do not commence close to the dental vessels and nerves, but a little above them in the thickness of the submucous tissue, nearly at the middle of the depth of the groove. The base of the follicle, as the organ grows larger, becomes nearer to the vessels and nerves.

The phenomena of the genesis and completion of the follicles have been minutely described by Robin and Magitot. The following are the principal steps of the process as observed by them:—A little before the first appearance of a follicle, that point of the submucous tissue at

---

(1) For an account of Raschkow's paper, see Nasmyth's "Historical Introduction."

(2) Ann. des Sciences Nat. Ser. iv., T. ix. (1858), and DENTAL REV., vol. ii., p. 401.

are the most complete. About this age, too, the ossifications

which it is about to originate becomes more opaque and vascular than the surrounding gelatiniform tissue. In the middle of the vascularity, which depends on a network of capillaries, a small, obscure, roundish mass appears. This is the bulb. When the bulb reaches a certain size, a dark, greyish band is formed round it. This represents the follicular wall. This band, after circumscribing the bulb, becomes raised above it, and eventually its free edge unites with itself, so as to convert the follicle into a closed sac. The cavity of the sac is soon divided into two equal parts, the upper of which is occupied by the organ of the enamel, which is formed immediately on the occlusion of the follicle—the lower by the bulb. At this time the bulb is in continuity with the surrounding tissue at its base, the rest of its periphery although immediately contiguous to the surrounding tissue, is easily detached, and has a smooth and very clearly defined surface. The bulb, from the time of its appearance, is formed of finely-granulated ovoid nuclei, separated from each other by a small quantity of granulated amorphous matter. At the time of its origin, the bulb of the incisor, or canine, is shaped like a cone, or more or less elliptical; that of the molar is blunt, more bulged, and broader at the base. The appearance of the follicular wall is accompanied by an increase of vascularity. The wall itself is traversed from its base to its summit by a capillary network, having regular polygonal meshes. The vessels and nerves of the bulbs are not developed until a little later, but always before the moment of the appearance of the first cells of origin of the dentine. The organ of the enamel has the appearance of a clear, transparent mass, situated between the inner surface of the wall and the progressing part of the bulb. It has no continuity of substance with the wall—it is only in contact with the surface of the bulb, from which a pale, white line separates it shortly after its appearance. This line consists of a continuous range of the cells of the enamel. As the follicle progresses, its wall becomes better marked, until it forms quite a resisting envelope, completely distinct from the neighbouring tissues. The base of the bulb, which is at first its broadest or almost its broadest part, becomes afterwards constricted and lengthened out to form the radicular portion of the bulb, which receives the vessels and nerves. The bulb at first consists only of ovoid or rounded nuclei with a little intervening amorphous matter. These nuclei are less transparent than the embryo-plastic nuclei. The follicular wall includes some embryo-plastic nuclei, a little firmly-granulated amorphous matter, and some laminated fibres, in the condition of fusiform bodies, or more completely developed. The organ of the enamel is a thin, gelatiniform bed, composed of star-studded fibro-plastic bodies, which are ramified and anastomosed. It soon exhibits, on its deep or bulbular surface, the continuous range of the cells of the enamel. They are prismatic and vertically arranged, and when seen as

begin on the edge of the first *Incisores*. The *Cuspidati* are not in the same circular line with the rest, but somewhat on the outside, making a projection there at this age, there not being sufficient room for them.(*w*)

In the sixth, or seventh month, the edges, or tips, of all these five substances have begun to ossify, and the first of them is a little advanced; and besides these, the pulp of the sixth Tooth has begun to be formed; it is situated in the tubercle of the Upper-Jaw, and under and on the inside of the Coronoid Process in the Lower-Jaw: so that at this age, in both Jaws, there are in all twenty Teeth begun to ossify, and the *stamina* of twenty-four.(*x*) They may be divided into the *Incisores*, *Cuspidati*, and *Molares*; for at this age there are no *Bicuspides*, the two last teeth in each side of both Jaws having all the characteristics, and answering all the purposes of the true *Molares* in the adult, though when these first *Molares* fall out, their places are taken by the *Bicuspides*.

The Teeth gradually advance in their ossification, and about the seventh, eighth, or ninth month after birth, the *Incisores* begin to cut or pass through the Gums; first, generally, in the Lower-Jaw. Before this time the ossifications in the third Grinder, or that which makes the first in the adult, are begun.

---

a whole under the microscope, appear like a clear band. It is on the surface of the bulb beneath the deep face of the enamel organ that the first cells of the dentine appear. Such is a sketch of the germs of the dental follicles, as seen by these observers. We shall hereafter notice the subsequent changes which are said by these observers to take place in the process of development. (1)]

(*w*) [An instance of an accordance in the embryonic state with the more general type.]

(*x*) [More recent observations refer the commencement of dentine to a somewhat earlier period. By the end of the fourth month, thin caps of dentine may be found surmounting the pulps of all the milk teeth, and a little later on the first permanent molars. According to Robin and Magitot, the first appearance of the tooth, properly so called, takes place between the eightieth and eighty-fifth day in the lower middle incisor.]

---

(1) Robin and Magitot, *op. cit.*

The *Cuspidatus* and *Molares* of the Fœtus are not formed so fast as the *Incisores;* they generally all appear nearly about the same time, viz. about the twentieth, or twenty-fourth month; however, the first Grinder is often more advanced within the socket than the *Cuspidatus,* and most commonly appears before it.(*y*)

These twenty are the only Teeth that are of use to the child from the seventh, eighth, or ninth month, till the twelfth or fourteenth year. These are called the Temporary or Milk Teeth, because they are all shed between the years of seven and fourteen, and are supplied by others.

OF THE CAUSE OF PAIN IN DENTITION.

These twenty Teeth, in cutting the Gum, give pain, and produce many other symptoms which often prove fatal to children in Dentition. It has been generally supposed that these symptoms arise from the Tooth's pressing upon the inside of the Gum, and working its way mechanically; but the following observations seem to be nearer the truth.

The Teeth when they begin to press against the Gum, irritate it, and commonly give pain. The Gums are then affected with heat, swelling, redness, and the other symptoms of inflammation. The Gum is not cut through by simple or mechanical pressure, but the irritation and consequent inflammation produces a thinning, or wasting of the Gum at this part: for it often happens that when an extraneous, or a dead substance, is contained in the body, that it produces a

---

(*y*) [The milk incisors usually are cut from the seventh to the ninth month after the birth; the first milk molars at about the twelfth month; the canines at about eighteen months; and the second milk molars at about twenty-four months. The first teeth cut are the central incisors of the lower jaw; these are followed by the same teeth in the upper; and, as a general rule, the lower tooth is cut earlier than the corresponding one above. According to Mr. Tomes, it is not until the child has attained the age of four and a half years that the temporary teeth are entirely perfected. He found in a child of four years and one month that the incisors were the only teeth fully formed. The fangs of the others were slightly deficient in length, and hollow at their extremities.]

destruction of the part between it, and that part of the skin which is nearest it, and seldom of the other parts, excepting those between it and the surface of a cavity opening externally, and that by no means so frequently. And in those cases there is an absorption of the solids, or of the part destroyed, not a melting down, or solution of them into Pus. The Teeth are to be looked upon as extraneous bodies, with respect to the Gum, and as such they irritate the inside of that part in the same manner as the Pus of an abscess, an exfoliation of a bone, or any other extraneous body; and therefore produce the same symptoms, excepting only the formation of matter.(z) If, therefore, these symptoms attend the cutting of the Teeth, there can be no doubt of the propriety of opening the way for them; nor is it ever, as far as I have observed, attended with any dangerous consequence.

OF THE FORMATION AND PROGRESS OF THE ADULT TEETH.

Having now considered the first formation, and the progress of the Temporary Teeth, we shall next describe the formation of those Teeth which are to serve through life.(a)

(z) [Any real analogy between the tooth and an extraneous body cannot be admitted. In a note on this section, Mr. Bell observes, "There can, indeed, be no doubt that the emancipation of the rising tooth is occasioned by absorption of the gum; but it is also probable that this absorption is increased, if not wholly produced, by the pressure of its edge on the horizontal surface of the tooth. It appears probable, therefore, that when, in consequence of the rapid elongation of the root, the crown of the tooth rises faster than this process for the removal of the containing parts goes on, an undue pressure takes place on the inside of the gum, and local inflammation, accompanied by much constitutional disturbance, is the result. The mere existence of the tooth in contact with the gum, 'as an extraneous body,' would not account for all this disturbance; for, after the gums are lanced, the tooth is still in contact with the soft parts; but because the pressure is thus taken off, the irritation immediately subsides. (1)"

(a) [According to the more commonly-received view, based on the

---

(1) 'Hunter's Works,' by Palmer, vol. ii., p. 35.

In this enquiry, to avoid confusion, I shall confine the description to the Teeth in the Lower-Jaw; for the only diffe-

---

observations of Goodsir, as has been already stated, the ten anterior permanent teeth in each jaw are formed in cavities of reserve behind the follicles of the ten milk teeth, whilst the three permanent molars on each side are formed in posterior cavities of reserve, formed by prolongations backwards of the primitive dental groove. About the fourteenth week, behind each milk follicle a small lunated recess, similar to an impression made by the nail, is said to be formed. These depressions escape the general adhesion of the sides of the dental groove, but by the closure of the latter they become converted into cavities, which are formed successively from before backwards, and ultimately become the sacs for the incisor, canine, and bicuspid teeth. These cavities elongate and sink into the substance of the gum, above and behind the upper milk follicles, below and behind the lower. A papilla is formed at the bottom of each, that for the central incisor appearing first, about the sixth month. Opercula, as is alleged in the case of the temporary teeth, are developed from the sides of the sac, dividing it into two portions, the lower of which encloses the papilla, which becomes the pulp of the permanent tooth. The upper and narrower part of the cavity becomes gradually obliterated by the adhesion of its sides. The permanent sac then adheres to the back of that for the temporary tooth. Both grow, and after a time it is found that each sac (the permanent and temporary) becomes lodged in a separate osseous recess or socket—that for the permanent tooth in the lower jaw being below and behind that for the temporary, in the upper jaw above and behind. A bony partition separates them. The permanent sac and its osseous recess present a pear-shape. The sac is connected with the gum by a solid membranous pedicle, which is contained in an osseous canal, which opens by an aperture on the edge of the jaw, behind the socket for the temporary tooth. The permanent tooth is formed in its sac in the same manner as the milk tooth, from which it is separated by a bony partition. In the rise of the permanent tooth through the gum, it presses on the bony partition and on the root of the milk tooth; these latter become absorbed under the influence of the pressure; the milk tooth, after the greater or less absorption of its fang, falls out, and the permanent tooth takes its place. A description of the formation of the posterior cavities of reserve, and of the papilla of the true molars, has already been given in the previous note on the dental pulps. (1) *Vide* note *v*. p.

Kölliker gives, on the whole, a similar account of the formation of the germs of the permanent teeth. He observes, that in the fourth

---

(1) *Vide* 'Quain's Anatomy,' by Professors Sharpey and Quain.

rence between those in the two Jaws, is in the time of their appearance, and generally it is later in the Upper-Jaw. Their formation and appearance proceeds not regularly from the first *Incisor* backwards to the *Dens Sapientiæ*, but it begins at two points on each side of both Jaws, viz. at the first *Incisor*, and at the first *Molaris*. The Teeth between these two points make a quicker progress than those behind.

The Pulp of the first adult *Incisor*, and of the first adult *Molaris* begin to appear in a Fœtus of seven or eight months, and five or six months after birth the ossification begins in them; soon after birth the Pulp of the second *Incisor* and *Cuspidatus* begin to be formed, and about eight or nine months afterwards they begin to ossify; about the fifth or sixth year the first *Bicuspis* appears; about the sixth or seventh the second *Bicuspis*, and the second *Molaris*; and about the twelfth, the third *Molaris* or *Dens Sapientiæ*.(*b*)

---

mouth, the cavities containing the papillæ of the temporary teeth become narrower and narrower, and at length are perfectly closed, " but in such a manner, that over each cavity, or tooth-sac, another small recess is formed as *cavity of reserve* for the twenty anterior permanent teeth, of which, even in the fifth month of fœtal life, the tooth germs are developed. At first the new cavities lie over the tooth sacs of the milk teeth, but they gradually move to the posterior side of them; and when the bony alveoli of the milk teeth appear, are received into small dilatations of them, which, in the incisor and canine teeth, become at last completely separate from the others; in the two first molars (bicuspides), on the other hand, open at the bottom of the alveoli of the milk teeth. The tooth sacs of all these teeth are drawn out at the apex in form of a solid cord, which extends either to the gum or on the two first molars (bicuspides), to the periosteum at the bottom of the two milk molars, and which has been incorrectly considered as being a conducting band (*gubernaculum*) of the teeth in their eruption. (1)"

Robin and Magitot deny that the follicles of the permanent teeth are developed in depressions on the surface of the mucous membrane of the gum.]

(*b*) [The papillæ of the first true molars appear as early as the sixteenth or seventeenth week according to Kölliker and Goodsir; somewhat earlier—the eighty-fifth to the ninety-fifth day—according to Robin and Magitot. Kölliker, as before noticed, states that the tooth germs of the

---

(1) Kölliker, *op. cit.*, p. 301.

The first five may be called the permanent Teeth: they differ from the temporary in having larger fangs. The permanent *Incisores* and *Cuspidati* are much thicker and broader, and the *Molares* are succeeded by *Bicuspides*, which are smaller, and have but one fang.

All these permanent or succeeding Teeth are formed in distinct *Alveoli* of their own; so that they do not fill up the old sockets of the temporary Teeth, but have their new *Alveoli* formed as the old *ones* decay.

---

twenty anterior permanent teeth are developed in the fifth month of fœtal life. Robin and Magitot say they originate some before and some after birth. According to Goodsir, the papilla of the second molar appears about the seventh month after birth, that of the third about the sixth year.

Calcification begins first in the anterior molar. The process is a little earlier in the teeth of the lower jaw than in the upper. The following is the order in which the process of calcification has been observed to take place in the permanent teeth of the upper jaw. First molar, five or six months; central incisor soon after; lateral incisor and canine, eight or nine months; two bicuspids, two years and over; second molar five or six years; third molar, or wisdom tooth, about twelve years.(1) Mr. Tomes gives the following description of the condition of the permanent teeth at the period when all the temporary teeth are perfected, but the first true molars have not cut the gums:—"The crowns of the permanent incisors, both of the upper and lower jaws, are perfected, excepting perhaps at that part where the enamel terminates. There the dull and chalk-like appearance, which that tissue presents when the development is progressing may be observed. The canines are still less advanced, while the crowns of the first bicuspids have not attained to more than two-thirds, and those of the second bicuspid not more than a third, of their ultimate lengths. The crowns of the first permanent molars are, as respects their external surface, fully developed; and the septa of dentine which extend across the base of the pulps marking out the several roots yet to be developed are fully pronounced. The second permanent molars are at present represented by about two-thirds of their crowns, and invested with a thin layer of partially-developed enamel. The positions of the pulps of the wisdom teeth are but faintly indicated by slight depressions in the bone posterior to the sockets, which contain the forming second molars. These marks may, however, at this period, be altogether wanting.(2)]

---

(1) Quain and Sharpey, *op. cit.*
(2) Tomes' System of Dental Surgery, p. 66.

The first *Incisor* is placed on the inside of the root of the corresponding temporary Tooth, and deeper in the Jaw. The second *Incisor* and the *Cuspidatus* begin to be formed on the inside, and somewhat under the temporary second *Incisor* and *Cuspidatus*. These three are all situated much in the same manner, with respect to the first set; but as they are larger, they are placed somewhat farther back in the circle of the Jaw.

The first *Bicuspis* is placed under, and somewhat farther back than the first temporary Grinder, or fourth Tooth of the child.

The second *Bicuspis* is placed immediately under the second temporary Grinder.

The second *Molaris* is situated in the lengthening tubercle in the Upper-Jaw, and directly under the Coronoid Process in the Lower.

The third *Molaris*, or *Dens Sapientiæ*, begins to form immediately under the Coronoid Process.

The first Adult *Molaris* comes to perfection, and cuts the Gum about the twelfth year of age, the second about the eighteenth, and the third, or *Dens Sapientiæ*, from the twentieth to the thirtieth: so that the *Incisores* and *Cuspidati* require about six or seven years, from their first appearance, to come to perfection; the *Bicuspides* about seven or eight; and the *Molares* about twelve.(c)

(c) [The eruption of the permanent teeth takes place a little earlier in the lower jaw than in the upper. The following tables record the results of the observations of Dr. Blake and Mr. Cartwright as the period of eruption:— (1)

BLAKE.	Years.	CARTWRIGHT.	Years.
Molar, First...............................	6½	Molar, First.......................	5 to 7
Incisors, Central.......................	7	Incisors, Central Inferior......	5 to 7
,, Lateral.........................	8	,, ,, Superior ...	6 to 8
Bicuspids, Anterior	9	,, Lateral.................	7 to 9
,, Posterior	10	Bicuspids, Anterior	8 to 10
Canines....................................	11 to 12	Canines	9 to 12
Molars, Second	12 to 13	Bicuspids, Posterior	10 to 12
,, Third (or Wisdom) ...	17 to 25	Molars, Second	12 to 14
		,, Third (Wisdom)	17 to 25

---

(1) From Quain's Anatomy, by Sharpey and Quain.

It sometimes happens that a third set of Teeth appears in very old people; when this does happen, it is in a very irregular manner, sometimes only one, at other times more, and now and

The following table gives the results obtained from an examination of 3,074 cases by Mr. S. Cartwright, jun. :—(1)

3074 Cases.	Between the 5th and 6th birthdays. Out of 170 children.	6th and 7th birthdays. Out of 340 children.	7th and 8th birthdays. Out of 496 children.	8th and 9th birthdays. Out of 530 children.	9th & 10th birthdays. Out of 454 children.	10th & 11th birthdays. Out of 322 children.	11th & 12th birthdays. Out of 303 children.	12th & 13th birthdays. Out of 203 children.	13th & 14th birthdays. Out of 140 children.	14th & 15th birthdays. Out of 86 children.	15th & 16th birthdays. Out of 30 children.
Lower Posterior Molars.	..	..	..	..	11	26	79	118	113	79	29
Upper Posterior Molars.	..	..	..	6	18	51	103	100	78	30	
Lower Anterior Molars.	48	199	479	524	453	322	303	203	140	86	30
Upper Anterior Molars.	34	182	472	524	453	322	303	203	140	86	30
Lower Posterior Bicuspids.	..	4	5	12	32	69	123	102	93	77	28
Upper Posterior Bicuspids.	1	2	3	16	51	110	166	144	122	79	30
Lower Anterior Bicuspids.	..	1	7	38	60	104	167	149	116	86	29
Upper Anterior Bicuspids.	..	3	19	85	143	199	231	175	133	86	30
Lower Cuspids.	..	..	..	7	40	98	166	159	120	83	30
Upper Cuspids.	..	..	..	8	20	48	112	136	115	79	29
Lower Incisors.	17	207	407	524	451	321	303	203	140	86	30
Upper Incisors.	5	52	180	459	435	318	303	203	140	86	30

(1) From Tomes' System of Dental Surgery.

then a complete set comes in both Jaws. I never saw an instance of this kind but once, and there two fore Teeth shot up in the Lower-jaw.

I should suppose that a new Alveolar Process must be also formed in such cases, in the same manner as in the production of the first and second sets of Teeth. From what I can learn, the age at which this happens is generally about seventy. From this circumstance, and another that sometimes happens to women at this age, it would appear that there is some effort in nature to renew the body at that period.

When this set of Teeth which happens so late in life, is not complete, especially where they come in one Jaw, and not in the other, they are rather hurtful than useful; for in that case we are obliged to pull them out, as they only wound the opposite Gum.(*d*)

THE MANNER IN WHICH A TOOTH IS FORMED.

The body of the Tooth is formed first, afterwards the Enamel and Fangs are added to it. All the Teeth are produced from a kind of pulpy substance, which is pretty firm in its texture, transparent, excepting at the surface, where it adheres to the

---

The order of eruption of the permanent teeth in the anthropoid apes presents a remarkable contrast to their sequence in man. Professor Owen remarks: " Both Chimpanzees and Orangs differ from the human subject in the order of development of the permanent series of teeth; the second molar (m 2) comes into place before either of the premolars has cut the gum, and the last molar (m 3) is acquired before the canine. We may well suppose that the larger grinders are earlier required by the frugivorous Chimpanzees and Orangs than by the higher organised omnivorous species with more numerous and varied resources, and probably one main condition of the earlier development of the canines and premolars in man may be their smaller relative size. (1)]

(*d*) [Many of these cases are doubtless instances of retarded eruption. A tooth which has been retained at length becomes exposed by the absorption of the superjacent gum.]

---

(1) Owen, Art. Odontology, Encyc. Brit. 8th Edit.

Jaw, and has at first the shape of the bodies of the Teeth which are to be formed from it. These pulpy substances are very vascular, they adhere only at one part to the Jaw, viz. at the bottom of the cavity which is to form the socket; and at that place their vessels enter; so that they are prominent, and somewhat loose in the bony cavity which lodges them.

They grow nearly as large as the body of the Tooth before the ossification begins, and increase a little for some time after the ossification is begun. (*e*) They are surrounded by a membrane, (*f*) which is not connected with them, excepting at their root or surface of adhesion. This membrane adheres by its outer surface all around the bony cavity in the Jaw, and also to the gum where it covers the Alveoli.

When the pulp is very young, as in the Fœtus of six or seven months, this membrane itself is pretty thick and gelatinous. We can examine it best in a new-born child, and we find it made up of two *lamellæ*, an external and internal: the external is soft and spongy, without any vessels; the other is much firmer, and extremely vascular, its vessels coming from those that are going to the pulp of the Tooth: it makes a kind of Capsula for the pulp and body of the Tooth. While the Tooth is within the gum, there is always a mucilaginous fluid, like the Sinovia in the joints, between this membrane and the pulp of the Tooth. (*g*)

(*e*) [According to Robin and Magitot the successive phases in the general evolution of the bulb are as follows:—1. A gradual increase of size until the period when the summit becomes covered by the cap of dentine. 2. Arrest at that time in general arrangement, but a continuation of growth in height, while the crown is simultaneously increased in width. 3. The coronary portion of the bulb ceases to grow larger, and radicular prolongations are produced from the base.]

(*f*) [The follicular wall. *Vide* Note v., p. 337.]

(*g*) [Hunter is mistaken in asserting that any portion of the follicular wall is nonvascular. Robin and Magitot observe that it is not probable that he refers to the enamel organ when he speaks of an internal lamella "firmer and extremely vascular." His description of a mucilaginous fluid like synovia between this membrane and the pulp of the tooth evidently points to the enamel organ. Kölliker describes the enamel organ as consisting of anastomosing stellate cells, or reticulated connective

When the Tooth cuts the gum, this membrane likewise is perforated; after which it begins to waste, and entirely gone by the time the Tooth is fully formed, for the lower part of the membrane continues to adhere to the neck of the Tooth, which has now risen as high as the edge of the gum. (*h*)

OF THE OSSIFICATION OF A TOOTH UPON THE PULP.(*i*)

The beginning of the ossification upon the pulp is by one point, or more, according to the kind of Tooth. In the *Incisores*

---

tissue, containing, in its interspaces a large quantity of fluid, rich in albumen and mucus. Robin and Magitot describe minutely the change in the enamel organ produced by putrefaction. They say that the enamel organ acquires the condition of a viscous fluid, which ropes like synovia, and is yellowish or reddish in colour. This matter consists of a finely-granulated fluid, containing some cells of enamel, and star-studded fibro-plastic bodies, with their fibrils broken off more or less near the nuclei. (1)]

(*h*) [The upper part of the tooth-sac is firmly united with the gum, and is spontaneously absorbed before the growing tooth. The remaining portion becomes closely applied to the fangs, and is converted into the dental periosteum.

(*i*) [Although since Hunter's time the microscope has been assiduously used to explain the method in which the earthy and animal substance of the tooth is deposited, the subject is confessedly so difficult that even at present observers are anything but unanimous in the opinions they hold as to the process. Kölliker describes the dental pulp as consisting of an "inner part rich in vessels, and subsequently also in nerves, and of an outer non-vascular portion. The latter is bounded by a delicate structureless membrane, the *membrana præformativa* of Raschkow, which is of no significance in the formation of the tooth; and beneath the membrane are cells, the dentinal cells, 0·016''' to 0·024''' long, and 0·002'' to 0·0045''' broad, with beautiful vesicular nuclei, and one or more distinct nucleoli, which are placed close to one another, like an epithelium upon the whole surface of the pulp; they are not, however, so sharply limited internally as an epithelium, and there is at least apparently a gradual transition by means of smaller cells between them and the parenchyma of the pulp. Nevertheless, in more vascular tooth pulps, a certain limitation arises from this, that the capillary loops of the vessels do not pass in between the cylindrical cells, but terminate close to each

---

(1) Kölliker, *Op. cit.*, p. 304. Robin and Magitot, *Op. cit.*

it is generally by three points, the middle one being the highest, and the first that begins to ossify. The *Cuspidatus* begins by

other underneath them, so that the designation of the layer of cells in question as the *dentinal membrane, membrana eboris*, appears to be warranted, especially as these cells really yield the dentine. The *inner parts* of the pulp consist throughout of a more granular or homogeneous, but subsequently more fibrous matrix, with numerous cell nuclei of a roundish or elongated form, which is to be regarded as a kind of connective tissue. At the period of ossification, vessels are developed in large numbers in the pulp, and the most numerous, perpendicularly disposed loops of capillaries of about 0·006''' are found principally upon the border of ossification. The nerves accompany the vessels, but are developed subsequently to them. Their number is likewise very considerable, and their distribution in the pulp similar to that in the fully developed teeth." (1) It is only the outermost epithelial-like cell-layer of the pulp, which is concerned in the formation of the dentine. Kölliker believes that the whole of the pulp is not converted from without inwards into dentinal cells and ossified, but that the pulp serves for the formation of the dentine simply by supporting the vessels which are necessary for the growth of the dentinal cells. Its decrease in the process of ossification, he thinks, is owing to the gradual absorption of its soft part, and not to its conversion into dentine. He holds that no other tissue but the dentinal cells contributes to the formation of dentine, and refers to some observations of Lent as explanatory of the process of transformation. In growing teeth which were macerated to disintegration in hydrochloric acid, Lent succeeded in isolating the cells in question with complete dentinal canals prolonged from them. A previous observation of Kölliker's had also partly led him to form the opinion that the dentinal cells became prolonged into the dentinal canals. He therefore concludes with Lent: 1. That " *the dentinal canals are direct processes of the whole dentinal cells,* which processes may send out subordinate branches, and anastomose by means of them. To all appearance a single cell seems in many cases to be sufficient to form an entire dentinal canal, or at least a very large part of one." 2. That " *the matrix of the dentine is not formed of the dentine cells,* but is a secretion of these cells and of the tooth pulp ; in other words, an intercellular substance." (2)

Professor Huxley believes that the tooth pulp is a dermic process bounded by a basement membrane having on its outer surface a layer of epithelium. (He also regards the follicular wall or capsule as an involution of the derm, which is lined with a basement membrane, having on

---

(1) Kölliker, *op. cit.*, p. 303-304.
(2) Kölliker, *op. cit.*, p. 307.

one point only; the *Bicuspis* by two, one external, which is the first and the highest, and the other internal. The *Molares*,

its inner surface a layer of epithelium. The two layers of epithelium, that of the pulp and that of the capsule, are in contact, and together form, according to Huxley, the so-called enamel organ.) The basement membrane of the pulp is identical with Nasmyth's "persistent membrane," and with the *membrana præformativa*. The dentine is formed beneath the basement membrane between it and the pulp. He holds that the dentine is *deposited* in the pulp, but that *no histological element of the pulp takes part in its formation*. (1)

According to Robin and Magitot, the tissue of the mass of the dental bulb or pulp is composed of ovoid nuclei sprinkled in great number through a slightly granular transparent homogeneous substance, and which are at a later period accompanied by laminated fibres, vessels, and nerves. These nuclei are analogous to embryo-plastic elements, but differ from them in several particulars, especially in having no nucleoli. The nuclei are generally disposed parallelly to each other, and their longest diameter is usually parallel to the vertical axis of the bulb. At the heart of the pulp and in the neighbourhood of the base near the point of continuity with the follicular wall, a certain number of nuclei become developed into fusiform or star-like fibroplastic bodies, and eventually into laminated fibres. The amorphous matter which occupies the intervals between the nuclei, extends on the surface of the bulb about the $\frac{1}{50}$ of a millimetre beyond the most superficial nuclei. Before the cells of the dentine originate the surface of this bed of amorphous matter becomes denser. (This denser layer is the membrana præformativa of Raschkow. It folds easily and is readily detached.) The whole bed of superficial amorphous matter ceases to exist at the point of juncture between the base of the bulb and the follicular wall. It is pale, very transparent, and destitute alike of nuclei and molecular granulation. In the depth of this bed the cells of the dentine originate, beneath that denser superficial layer (membrana præformativa, Raschkow) which is capable of detachment, and which continues until the time when the cells change into the condition of ivory. These observers believe that the dentinal cells originating in the superficial bed of amorphous matter, grow by individual genesis, and are not the product of any other organ. They are oblong, nucleated, disposed in rows parallel to each other, and in consequence of their relative approximation, they assume a prismatic form, with four, five, or six sides. They are connected by their central

---

(1) *Vide* Huxley on the Development of the Teeth. 'Microscopic Journal,' 1853, p. 149.

either in a child, or an adult, begin by four or five ossifications, one on each point, the external always the first. Where the

ends with the tissue of the bulb, by their peripherical ends with the small quantity of bulb matter which extends beyond them. These dentinal cells are transformed into ivory: the change extends from the periphery of the cell to the central end. " When the alteration into the solid condition is once effected, the cell becomes for ever unrecognisable, the body having gradually changed from the condition of a soft substance to that of ivory, and the nucleus becomes atrophied during the general invasion." Robin and Magitot therefore are of opinion that the *matrix of the dentine is formed by the calcification of the dentinal cells*. (1)

With regard to the formation of the cement, according to Kölliker, it proceeds from that part of the tooth sac which is situated between the pulp and the enamel organ. It commences as soon as the fang begins to be formed. " At this period the tooth sac becomes elongated at its lower part, is closely applied to the developing fang, and, from its rich network of vessels, furnishes—as the periosteum does during the growth in thickness of the bones—a soft blastema, in which nucleated cells become developed, and which then immediately ossifies. Accordingly the cement is not formed by the ossification of the tooth sac itself." Upon the dentine the membrana præformativa becomes covered by the deposition of the cement, and is then no longer to be recognised. On the enamel, on the contrary, it remains as Nasmyth's membrane or the cuticle of the enamel. (2)

Professor Hannover, of Copenhagen, entertains some original views on the subject of the formation of the dentine and cement. According to this observer the dental follicle, in addition to the germs for the production of the different dental tissues, consists of what he terms the *membrana intermedia*, a membrane which separates the cement from both the enamel and the dentine. On the crown of the tooth, the enamel cells are attached to the inner surface of this membrane, and in those animals in which cement exists on the crown, it there separates the enamel from the cement. At the root of the tooth, where there is no enamel, the membrane lies between the dental germ and the cement germ. Dr. Hannover thinks that in the fully-formed tooth this *membrana intermedia* is metamorphosed into a *stratum intermedium*,—the " granular layer " of other writers. On the crown, Dr. Hannover believes that the so-called " cuticle of the enamel" is subsequently formed by this membrana intermedia. According to this observer, the dentinal tubules are formed by the dentinal cells, which arrange themselves in

---

(1) Robin and Magitot, *op. cit.*
(2) Kölliker, *op. cit.*, p. 307.

Teeth began to ossify at one point only, that ossification gradually advances till the Tooth is entirely completed; but if there are more than one point of ossification, each ossification increases till their basis come in contact with one another, and there all unite into one; after which they advance in growth as one ossification.

The ossifications in their progress become thicker and thicker where they first began, but increase faster on the edges of the Teeth; so as thence to become more and more hollow, and the cavity becomes deeper. As the ossification advances, it gradually surrounds the Pulp till the whole is covered by bone, excepting the under surface: and while the ossifications advance, that part of the Pulp which is covered by bone is always more vascular than the part which is not yet covered.

The adhesion of the pulp to the new-formed Tooth, or bone, is very slight, for it can always be separated from it without any apparent violence, nor are there any vessels going from the one to the other; the place, however, where it is most strongly attached is round the edge of the bony part, which is the last part formed. When the bone has covered all the Pulp, it begins to contract a little, and becomes somewhat rounded, making that part of the Tooth which is called the neck; and from this place the fangs begin. When the fangs form, they

---

rows on the surface of the pulp and send forth prolongations. The prolongation of the anterior end of one cell coalesces with the posterior prolongation of another; and in this way the tube is formed, which becomes afterwards calcified. Dr. Hannover has minutely described the formation of cement, by the production of cartilage cells in what he calls the cement germ, and subsequent ossification. (3) Robin and Magitot, however, assert that he has described the proper tissue of the organ of the enamel as being the organ of the cement, and that what Hannover terms the "membrana intermedia" is only a modification of the texture of the organ of the enamel at its internal face.]

---

(3) *Vide* 'Medico-Chirurgical Review,' 1857.

push up the bodies of the Teeth through the sockets, which waste, and afterwards through the Gum, which also wastes, as has been explained upon the cutting of the Teeth; for before this time the rising of the Teeth is scarce observable, as the Pulp was at first nearly of the size of the body of the Tooth itself, and wasted nearly in proportion to the increase of the whole ossification

The Pulp has originally no process answering to the fang; but as the cavity in the body of the Tooth is filled up by the ossification, the pulp is lengthened into a fang. The fang grows in length, and rises higher and higher in the socket, till the whole body of the Tooth is pushed out. The socket, at the same time, contracts at its bottom, and grasps the neck, or beginning fang, adheres to it, and rises with it, which contraction is continued through the whole length of the socket as the fang rises; or the socket which contained the body of the Tooth, being too large for the fang, is wasted or absorbed into the constitution, and a new Alveolar portion is raised with the fang; whence in reality the fang does not sink, or descend into the Jaw. Both in the body, and in the fang of a growing Tooth, the extreme edge of the ossification is so thin, transparent, and flexible, that it would appear rather to be horny than bony, very much like the mouth or edge of the shell of a snail when it is growing: and indeed it would seem to grow much in the same manner, and the ossified part of a Tooth would seem to have much the same connection with the pulp as a snail has with its shell.

As the Tooth grows, its cavity becomes gradually smaller, especially towards the point of the fang. In tracing the formation of the fang of a Tooth, we hitherto have been supposing it to be single, but where there are two, or more, it is somewhat different, and more complicated.

When the body of a Molares is formed, there is but one general cavity in the body of the Tooth, from the brim of which the ossification is to shoot, so as to form two or three fangs. If two only, then the opposite parts of the brim of the cavity of the Tooth shoot across where the Pulp adheres to the Jaw, meet in the middle, and thereby divide the mouth of the cavity into two openings; and from the edges of these two openings the two fangs grow.

We often find that a distinct ossification begins in the middle of the general cavity upon the root of the Pulp, and two processes coming from the opposite edges of the bony shell join it; which answers the same purpose.

When there are three fangs, we see three processes coming from so many points of the brim of the cavity, which meet in the center, and divide the whole into three openings; and from these are formed the three fangs. We often find the fangs forked at their points, especially in the *Bicuspides*. In this case, the sides of the fang as it grows, come close together in the middle, making a longitudinal groove on the outside; and this union of the opposite sides divides the mouth of the growing fang into two orifices, from which the two points are formed.

By the observations which I have made in unravelling the texture of the Teeth, when softened by an acid, and from observing the disposition of the red parts in the Tooth of growing animals, interruptedly fed with madder, I find that the bony part of a Tooth is formed of *Lamellæ*, placed one within another. The outer *Lamella* is the first formed, and is the shortest: the more internal *Lamellæ* lengthen gradually towards the fang, by which means, in proportion as the Tooth grows longer, its cavity grows smaller, and its sides grow thicker.

How the earthy and animal substance of the Tooth is deposited on the surface of the pulp is not perhaps to be explained.

OF THE FORMATION OF THE ENAMEL.

In speaking of the Enamel we postponed treating of its Formation, till it could be more clearly understood; and now we shall previously describe some parts which we apprehend to be subservient to its formation, much in the same manner as the Pulp is to the body of the Tooth.

From its situation, and from the manner in which the Teeth grow, one would imagine that the Enamel is first formed; but the bony part begins first, and very soon after the Enamel is formed upon it. There is another pulpy substance opposite to that which we have described, (*j*) it adheres to the inside of the

---

(*j*) We have seen (note *g*) that Hunter has already alluded to the so-called organ of the enamel; he here describes it minutely. There is a great difference of opinion amongst writers upon the subject of the formation of the enamel, and on the function of the so-called enamel organ.

Kölliker describes the enamel organ, *organon adamantinæ*, as embracing with its inner concave surface the tooth pulp in its entire extent and as being connected at its outer side with the tooth sac, but in such a manner that it possesses a very small free border at the base of the tooth germ. Its principal mass consists of anastomosing stellate cells, or reticulated connective tissue, enclosing in its interspaces a large quantity of albuminous fluid. "This gelatinous areolar tissue is thickest immediately before the commencement, and in the first stages of ossification. Thus, in the fifth and sixth months, it is $\frac{4}{10}$ to $\frac{2}{3}$ of a Vienna line; in the newly-born infant, on the other hand, only 0·16″ to 0·2″. At this period it also possesses vessels in its outer third, and its network has become metamorphosed into true connective tissue. Upon the inner side of the spongy tissue of the enamel organ, there is situated the so-called *enamel membrane, membrana adamantinæ* (Raschkow) a genuine cylinder epithelium, with cells which measure 0·012‴ in length, 0·002‴ in breadth, are finely granular and delicate, and contain elongated, round nuclei, often situated at the apices of the cells."

Schwann, and most other authors, have assumed that the enamel fibres are nothing else than the ossified cells of the enamel membrane. Professor Huxley, on the contrary, asserts that this is impossible, because the enamel in all stages of its development is covered by the membrana præformativa of the dental pulp, and is separated from it by the enamel membrane. Therefore he asserts that the enamel is formed independently of the enamel membrane, and beneath the membrana præformativa, which afterwards is converted into the cuticle of the enamel, or Nasmyth's membrane. Huxley's observations have been repeated, and to a great extent confirmed, by Lent, who finds that undoubtedly a delicate structureless membrane may at all times be separated from the surface of the developing enamel when treated with dilute acids, and that this membrane, whilst the dentine is not fully

Capsula, where the Gum is joined to it, and its opposite surface lies in contact with the basis of the above described Pulp, and formed, is continued into the membrane præformativa of the tooth-pulp. Kölliker observes that if we assume the correctness of these observations, the following seem to be the only admissable explanations of the mode in which the enamel is formed.

1. "The enamel fibres are produced by a secretion of the cells of the enamel membrane, which penetrates the *membrana præformativa* in a fluid condition, but hardens and ossifies beneath it.

Or 2. That "the enamel fibres are formed by the dentine from an exudation furnished by the dentinal canals." This second supposition, he adds, leaves the facts of the formation of regular fibres in the enamel, and its growth in thickness by means of the apposition of new layers upon its outer surface wholly unexplained. The former assumption although more probable is also beset with difficulties. (1)

Robin and Magitot distinctly confirm the observations of Huxley and Lent. They describe the organ of the enamel as consisting of fibroplastic bodies and of amorphous matter interposed between the fibroplastic bodies. The tissue of the organ of the enamel they assert is absolutely without blood capillaries and nevous filaments. It is surrounded on all sides by an epithelial layer, consisting of a continuous single range of cells. On the bulb face of the organ these cells are prismatic (membrana adamantinæ, Raschkow); these cells, however, are not transformed into enamel fibres, for the prismatic epithelial cells are always separated from the enamel fibres by the pellicle styled the membrana præformativa. The prisms of the enamel, according to these authors, originate by autogenesis at the surface of the ivory, and from the first, have individually the density, form, consistency, and brittleness which they always exhibit. "Their development is nothing more than the molecular phenomenon of which their appearance is the result, which, continuing to operate at the extremity opposite to the ivory, causes their progressive elongation." (2)

Mr. Tomes, however, has found reason to doubt the correctness of Professor Huxley's observations. He asserts that proceeding in the mode of examination adopted by Huxley and Lent, "No difficulty attended the production of the membrane, but the columns of the enamel pulp (*i. e.*, the prismatic cells of the enamel membrane) were found at many points adherent, and their continuity with the (enamel) fibres could be distinctly traced. Again, the detached columns adhered in bundles to each other by the ends which approached the enamel, and many of the

---

(1) Kölliker, *op. cit.* p. 304-6.
(2) Robin and Magitot, *op.cit.*

afterwards with the new formed basis of the Tooth: whatever eminences or cavities the one has, the other has the same, but reversed, so that they are moulded exactly to each other.

In the *Incisores* it lies in contact not with the sharper cutting edge of the Pulp, or Tooth, but against the hollowed inside of the Tooth; and in the *Molares* it is placed directly against their base, like a Tooth of the opposite Jaw. It is thinner than the other Pulp, and decreases in proportion as the Teeth advance. It does not seem to be very vascular. The best time for examining it is in a Fœtus of seven or eight months old.

In the granivorous animal, such as the horse, cow, &c., whose Teeth have the Enamel intermixed with the bony part, and whose Teeth, when forming, have as many interstices as there are continuations of the Enamel, we find processes from the Pulp passing down into those interstices as far as the Pulp which the Tooth is formed from, and there coming into contact with it.

After the points of the first described Pulp are begun to

---

columns were terminated by delicate processes, which must, at the time of separation, have been withdrawn from the interior of the partially-calcified fibres, and must consequently have passed through the membrane which is supposed to separate the two tissues." (1) Mr. Tomes concludes from preparations in his possession "that the columns of the enamel organ must be regarded as subservient to the development of the fibres, the conversion of the one into the other taking place in the following manner:—The proximal end of the column becomes calcined, not uniformly throughout its thickness, but the outer surface or sheath first receives the salts of lime, and at the same time the columns become united laterally. At this point, that is, at the extreme margin of calcification, the columns readily separate from the fibres, and leave a surface which, when looked upon directly, has the appearance of a membrane, the reticulate character of which is due to the withdrawal of the central portion of the calcifying column," this central portion being the process described above. (2)]

---

(1) Tomes *op. cit.*, p. 267.
(2) *Op. cit.*, p. 270.

ossify, a thin covering of enamel is spread over them, which increases in thickness till some time before the Tooth begins to cut the Gum.

The Enamel appears to be secreted from the Pulp above described, and perhaps from the Capsula which incloses the body of the Tooth. That it is from the Pulp and Capsula seems evident in the horse, ass, ox, sheep, &c. therefore we have little reason to doubt of it in the human species. It is a calcareous earth, probably dissolved in the juices of our body, and thrown out from those parts which act here as a gland. After it is secreted, the earth is attracted by the bony part of the Tooth, which is already formed; and upon that surface it crystallizes.

The operation is similar to the formation of the shell of the egg, the stone in the kidneys and bladder, and the gall stone. This accounts for the striated crystallized appearance which the Enamel has when broken, and also for the direction of these *Striæ*. *

The Enamel is thicker at the points and basis than at the neck of the Teeth, which may be easily accounted for from its manner of formation; for if we suppose it to be always secreting, and laid equally over the whole surface, as the Tooth grows, the first formed will be the thickest; and the neck of the Tooth, which is the last formed part inclosed in this Capsula, must have the thinnest coat; and the fang, where the *Periosteum* adheres, and leaves no vacant space, will have none of the Enamel.

At its first formation it is not very hard; for by exposing a very young Tooth to the air, the Enamel cracks, and looks rough: but by the time that the Teeth cut the Gum, the Enamel seems to be as hard as ever it is afterward; so that the air seems to have no effect in hardening it.

---

* The Author has made many experiments on the formation of different *Calculi*, and finds they are formed by crystallization, which were communicated to his brother, and taught by him to his pupils in 1761, and which he proposes to give to the Public as soon as his time will permit.

OF THE MANNER OF SHEDDING OF TEETH.

An opinion has commonly prevailed, that the first set of Teeth are pushed out by the second; this, however, is very far from being the case: and were it so, it would be attended with a very obvious inconvenience; for, were a Tooth pushed out by one underneath, that Tooth must rise in proportion to the growth of the succeeding one, and stand in the same proportion above the rest. But this circumstance never happens: neither can it; for, the succeeding Teeth are formed in new and distinct sockets, and generally the *Incisores* and the *Cuspidati* of the second set are situated on the inside of the corresponding teeth of the first set; and we find, that in proportion to the growth of the succeeding Teeth, the fangs of the first set decay, till the whole of the fang is so far destroyed, that nothing remains but the neck, or that part of the fang to which the Gum adheres, and then the least force pushes the Tooth out. It would be very natural to suppose, that this was owing to a constant pressure from the rising Teeth against the fangs or sockets of the first set: but it is not so; for, the new Alveoli rise with the new Teeth, and the old Alveoli decay in proportion as the fangs of the old Teeth decay, and when the first set falls out, the succeeding Teeth are so far from having destroyed, by their pressure, the parts against which they might be supposed to push, that they are still inclosed, and covered by a complete bony socket. From this we see, that the change is not produced by a mechanical pressure, but is a particular process in the animal œconomy.

I have seen two or three Jaws where the second temporary Grinders were shedding in the common way, without any Tooth underneath; and in one Jaw, where both the Grinders were shedding, I met with the same circumstance.

A remarkable instance of this sort occurred to me in a lady who desired me to look at a loose Tooth, which I found was the last temporary Tooth not yet shed. I desired that it might be drawn out, and told her it was of no use, and could not by any art be fixed, as it was one of the Teeth that is naturally shed, and that another might come in its place: however she was disappointed.

These cases prove evidently, that in shedding, the first Teeth are not pushed out by the second set, but that they grow loose, and fall out of their own accord. That the succeeding Teeth have some influence on the shedding of the temporary set is proved by those very cases; since in one of the first mentioned the person was above twenty years of age, and in the other the lady was thirty; and it is reasonable to believe, that the shedding of these Teeth was so late in those instances, from the want of the influence, whatever it is, of the new Teeth. When the *Incisores* and *Cuspidati* of the new set are a little advanced, but long before they appear through their bony sockets, there are small holes leading to them on the inside, or behind the temporary sockets and Teeth; and these holes grow larger and larger, till at last the body of the Tooth passes quite through them.(*k*)

OF THE GROWTH OF THE TWO JAWS.

As a knowledge of the manner in which the two Jaws grow, will lead to the better understanding the shedding of the Teeth; and as the Jaws seem to differ in their manner of growing, from other bones, and also vary according to the age, it will be here proper to give some account of their growth.

In a Fœtus three or four months old, we have described the

(*k*) [The process by which the fang of the milk-tooth becomes absorbed is not at present understood. The partition which separates the alveolus of the milk from the permanent tooth becomes removed, so that the latter comes to lie under the former, and as the permanent tooth advances, the milk, already loosened by the absorption of its fang from below upwards, gives place to it. The small holes erroneously stated in the text to be the commencement of openings through which the permanent teeth are to pass are the foramin through which the cords pass connecting the permanent tooth-sacs with the gum. (*Vide* Note, a. p. 313.)]

marks of four or five Teeth, which occupy the whole length of the Upper-Jaw, and all that part of the Lower which lies before the Coronoid Process, for the fifth Tooth is somewhat under that process.

These five marks become larger, and the Jaw bones of course increase in all directions, but more considerably backwards; for in a Fœtus of seven or eight months, the marks of six Teeth in each side of both Jaws are to be observed, and the sixth seems to be in the place where the fifth was; so that in these last four months the Jaw has grown in all directions in proportion to the increased size of the Teeth, and besides has lengthened itself at its posterior end as much as the whole breadth of the socket of that sixth Tooth.

The Jaw still increases in all points till twelve months after birth, when the bodies of all the six Teeth are pretty well formed; but it never after increases in length between the symphysis and the sixth Tooth; and from this time too, the Alveolar Process, which makes the anterior part of the arches of both Jaws, never becomes a section of a larger circle, whence the lower part of a child's face is flatter, or not so projecting forwards as in the adult.

After this time the Jaws lengthen only at their posterior ends; so that the sixth Tooth, which was under the Coronoid Process in the Lower-Jaw, and in the tubercles of the Upper-Jaw of the Fœtus, is at last, *viz.* in the eighth or ninth year, placed before these parts; and then the seventh Tooth appears in the place which the sixth occupied, with respect to the Coronoid Process, and tubercle; and about the twelfth or fourteenth year, the eighth Tooth is situated where the seventh was placed. At the age of eighteen, or twenty, the eighth Tooth is found before the Coronoid Process in the Lower-Jaw, and under, or somewhat before the tubercle in the Upper-Jaw, which tubercle is no more than a succession of sockets for the Teeth till they are completely formed.

In a young child the cavity in the temporal bone for the articulation of the Jaw is nearly in a line with the Gums of the

Upper-Jaw ; and for this reason the Condyle of the Lower-Jaw is nearly in the same line ; but afterwards, by the addition of the Alveolar Process and Teeth, the line of the Gums in the Upper-Jaw descends considerably below the articular cavity ; and for that reason the Condyloid Process is then lengthened in the same proportion.

In old people who have lost all their Teeth, the articulation comes again into the same line with the Gums of the Upper-Jaw; but in the Lower-Jaw, the Condyles cannot be diminished again for accommodating it to the Upper, so that it necessarily projects beyond the Gums of the Upper-Jaw at the fore part. When the mouth is shut, the projection of the Jaw at the chin, fits the two Jaws to each other at that place where the Grinders were situated, and where the strength of mastication lies; for if the chin was not further from the center of motion than the Gums of the Upper-Jaw, at the fore part, the Jaws, in such people as have lost all their Teeth, would meet in a point at the fore part, like a pair of pinchers, and be at a considerable distance behind.

THE REASON FOR THE SHEDDING OF THE TEETH.

As the shedding of the Teeth is a very singular process in the animal œconomy, many reasons have been asssigned for it ; but these reasons have not carried along with them that conviction which is desired. Authors have not fully considered the appearances which naturally explain themselves ; nor have they considered the advantages necessarily arising from the size and construction of only such a number as the first set; nor have they considered fully the disadvantages that such size and construction would have, if continued when it is necessary to have a greater number, which is the case with the adult.

We shall consider these advantages in a child where the shedding Teeth are all completely formed, which will be setting them in the clearest point of light ; and also, the disadvantages that would occur, if in the adult these were not changed for another set somewhat different.

If the child had been so contrived, as not to have required Teeth till the time of the second set's appearing, there would have been no occasion for a new set: but the Jaw-bones being considerably smaller in children than in adults, and it being necessary that they should have two Grinders, there is not room for *Incisores* and *Cuspidati* of sufficient size to serve through life; and the first formed Grinders having necessarily too small fangs, and the Jaw increasing at the back part only, these two Grinders would have been protruded too far forwards, and at too great a distance from the center of motion. This variation in the size of the Teeth is likewise a reason why the second set are not formed in the sockets of the first; and why the old sockets are destroyed.

These circumstances with regard to the shedding of the Teeth, contradict the notion of the second set being made broader and thicker, by the resistance they meet with in pushing out the first. For were we, on a partial view of the subject, to admit the supposition, the *Bicuspides* would effectually overturn our hypothesis; because here the second set are much smaller than the first, and yet the resistance would be greater to them than to the *Incisores*.

From the manner in which the Teeth are shed, it is evident that drawing a temporary Tooth, for the easier protrusion of the one underneath, will be of no great service; for in general it falls out before the other can touch it. But it is of much more service to pull out the neighbouring, or adjacent temporary Tooth; for we must be convinced by what has been advanced with regard to the changes in size, that excepting the whole were to shed at the same time, or the order of shedding, viz. from before backwards, were to be inverted, that the second set of *Incisores* and *Cuspidati* must be pinched in room, till the Grinders are also shed: and therefore we find it often of use to draw a temporary Tooth, that is placed further back; and it would, perhaps, be right upon the whole, always to draw, at least the first Grinder, and, perhaps, some time after, the second Grinder also. (*l*)

OF THE CAVITY FILLING UP AS THE TEETH WEAR DOWN.

A Tooth very often wears down so low, that its cavity would be exposed, if no other alteration were produced in it. To prevent this, nature has taken care that the bottom part of the cavity should be filled up by new matter, in proportion as the surface of the Teeth is worn down.(*m*) This new matter may be easily known from the old; for when a Tooth has been worn down almost to the neck, a spot may always be seen in the middle,

---

(*l*) [The practice here recommended, "always to draw at least the first grinder" of the temporary set has been too much followed by the interested or ignorant, who have readily shielded their malpractice under the authority of Hunter's name. The early extraction of any of the temporary teeth to make room for the permanent ones is rarely necessary, and it is on all accounts to be deprecated, unless the peculiar circumstances of the case imperatively call for it. But the removal of the large molar teeth in the child, in *anticipation* of a future deficiency of room, is so obviously uncalled for, and such a wanton interference with the usual process of nature, that we cannot but wonder at its being proposed as a general rule, even were there no positive evils to be apprehended from it: but this is not all; not only does the premature removal of the temporary molars endanger the perfect formation of the bicuspids which succeed them, by the forcible laceration of the connecting cord before described, but, if it take place before the permanent teeth are ready to fall into their ultimate situation, the jaw will contract as the child grows, and the second set of teeth will be forced into an irregular position, from permanent want of room. These arguments hold good against the too early removal of any of the deciduous teeth.] T. Bell.

(*m*) [The pulp cavity becomes filled with a kind of irregular dentine, and is occasionally completely obliterated by it. In caries the tooth is sometimes found strengthened by additions of denture on the wall of the pulp cavity.(1) Mr. Bell notices that filling up of the internal cavity of the tooth is frequently observed in sailors who have lived much on hard biscuit.]

---

(1) Vide Tomes op. cit., p. 327.

which is more transparent, and at the same time of a darker colour, (occasioned, in some measure, by the dark cavity under it) and is generally softer than the other. Any person may be convinced of the truth of these observations, by taking two Teeth of the same class, but of very different ages, one just completely formed, the other worn down almost to its neck. In the last he will observe the dark spot in the center; and if as much is cut off from the complete Tooth as hath been worn off the old one, the cavity of the young Tooth will be found cut through; and on examining the other, its cavity will be found filled up below that surface. Now this observation contradicts the idea of the hole leading into the cavity of the Tooth being closed up; and what is still a further proof of it, I have been able to inject vessels in the cavities of the Teeth in very old people when the Alveolar Process has been gone, and the Teeth very loose in the Gum.

Old people are often found to have very good sets of Teeth, only pretty much worn down. The reason of this is, that such people never had any disorder in their Teeth, or Alveolar Processes, sufficient to occasion the falling of one Tooth. For if by accident one Tooth is lost, the rest will necessarily fail in some degree, even though they are sound, and likely to remain so, had not this accident happened; and this weakening cause is greater, in proportion to the number that are lost. From this observation, we see that the Teeth support one another.

OF THE CONTINUAL GROWTH OF THE TEETH.

It has been asserted that the Teeth are continually growing, and that the abrasion is sufficient to keep them always of the same length; but we find that they grow at once to their full length, and that they gradually wear down afterwards; and that there is not even the appearance of their continuing to grow. The Teeth would probably project a little farther out of the Gum, if they were not opposed by those in the opposite Jaw; for in young people, who had lost a Tooth before the rest had come to their full length, I have seen the opposite Tooth project a little beyond the rest, before they were at all worn down. It may be further observed, that when a Tooth is lost, the opposite one may project from the disposition of the Alveolar Process to rise higher, and fill up at the bottom of the sockets; and the want of that natural pressure seems to give that disposition to these processes, which is best illustrated in those Teeth which are formed deeper in their sockets than usual. As a proof that the Teeth continue growing, it has been said that the space of a fallen Tooth is almost filled up by the increased thickness of the two adjacent Teeth, and the lengthening of that which is opposite. There is an evident fallacy in the case; either the observations have been made upon such Jaws as above described, or the appearances have not been examined with sufficient accuracy; for when the space appears to have become narrow by the approximation of the two adjacent Teeth, it is not owing to any increase of their breadth, but to their moving from that side where they are well supported, to the other side, where they are not. For this reason they get an inclined direction; and I observe it extends to the several adjacent Teeth in a proportional less degree, and affects those which are behind, more than those which are before the vacant space.

In the Lower-Jaw the back Teeth are not fixed perpendicularly, but all inclined forward; and the depression of the Jaw increases this position: the action of the Teeth, when thrown out of the perpendicular, has also a tendency to increase that oblique direction, as a pair of scissors, in cutting, pushes every thing forward, or from the center of motion: therefore this alteration, I think, is most commonly observable in the Lower-Jaw.

And that Teeth are not actually always growing in breadth, must be obvious to every person who considers, that in many people, through life, the Teeth stand so wide from each other, that there are considerable spaces between them; which could not be the case if they were always growing in thickness.

We might add too, that according to the hypothesis, that *Dens Sapientiæ* should grow to an enormous size backward, because there it has no check from pressure; and in people where the *Dens Sapientiæ* is wanting in one Jaw, which is very common, it should grow to an uncommon length in the opposite Jaw, for the same reason. But neither of these things happens.

I need hardly take notice, that when a Tooth has lost its opposite, it will in time become really so much longer than the rest, as the others grow shorter by abrasion; and I observe that the Tooth which is opposite to the empty space, becomes in time not only longer for the above-mentioned reason, but more pointed. The apex falls into the void space, and the two sides are rubbed away against the sides of the two approaching Teeth next to that space.

The manner of their formation likewise shews that Teeth cannot grow beyond a certain limited size. To illustrate this I may observe, that I have often, in the dead body of adults, found the left *Cuspidatus* of the Upper-Jaw, with its points scarcely protruding out of the Alveolar Process, though the Tooth was completely formed, and longer than the other by the whole point, which in that other was worn away. This Tooth, at its first formation, had been deeper in the Jaw than what is common; and after it had grown to the ordinary size, it grew no longer, though it had not the resistance of the opposite Teeth to set bounds to its increase: yet commonly in these cases the Tooth continues to project further and further through the Gum; though this is not owing to its growing longer, but to the socket filling up behind it, and thereby continuing to push it out by slow degrees.

OF THE SENSIBILITY OF TEETH.

The Teeth would seem to be very sensible; for they appear to be subject to great pain, and are easily and quickly affected by either heat or cold.

We may presume that the bony substance itself, is not capable of conveying sensations to the mind, because it is worn down in mastication, and occasionally worked upon by operators in living bodies, without giving any sensation of pain in the part itself. (*n*)

In the cavity of a Tooth it is well known that there is exquisite sensibility; and it is likewise believed, that this is owing to the nerve in that cavity. This nerve would seem to be more sensible than nerves are in common; as we do not observe the same violent effects from any other nerve in the body being exposed either by wound, or sore, as we do from the exposure of the nerve of a Tooth. Perhaps the reason of the intenseness, as well as the quickness of the sense of heat and cold in the Teeth, may be owing to their communicating these to the nerve sooner than any other part of the body.

OF SUPERNUMERARY TEETH.

We often meet with Supernumerary Teeth; and this, as well as some other variations, happens oftener in the Upper than in the Lower-Jaw, and, I believe, always in the *Incisores* and *Cuspidati*. I have only met with one instance of this sort,

---

(*n*) [That the bony substance of the teeth "is itself capable of conveying sensation to the mind" is, notwithstanding the author's presumption to the contrary, easily proved. Whence otherwise arises that acute sensation so commonly felt when the neck of a tooth is touched with the nail or any sharp instrument, or when a portion of the enamel only is broken from the surface of a tooth? In the latter case every other part of the tooth may be touched without any sensation being produced; but as soon as the instrument comes in contact with the denuded portion of the bone, a painful acute sensation is instantly perceived.] T. Bell.

and it was in the Upper-Jaw of a child about nine months old: there were the bodies of two Teeth, in shape like the *Cuspidati*, placed directly behind the bodies of the two first permanent *Incisores*; so that there were three Teeth in a row placed behind one another, viz., the temporary *Incisor*, the body of the permanent *Incisor*, and that supernumerary Tooth. The most remarkable circumstance was, that these Supernumerary Teeth were inverted, their points being turned upwards, and bended by the bone which was above them not giving way to their growth, as the Alveolar process does. (*o*)

It often happens that the *Incisores* and *Cuspidati*, in the Upper-Jaw especially, are so irregularly placed, as to give the appearance of a double row. I once saw a remarkable instance of this in a boy; the second *Incisor* in each side was placed farther back than what is common, and the *Cuspidatus* and first *Incisor* closer together, than if the second *Incisor* had been directly between them; so that the appearance gave an idea of a second row of teeth.

(*o*) [Mr. Tomes observes that "Supernumerary teeth may spring up during the second dentition in any part of the alveolar arch, and the forms of such teeth may either resemble those of special members of the normal series, or they may deviate from each of the recognised forms, and assume a somewhat irregular conical shape, sufficiently characteristic in itself to be at once recognised as that of a supernumerary tooth." (1) He mentions a case in which there were as many as four supernumerary teeth forming a group with the upper incisors and canines. Sometimes supernumerary teeth are not distinguishable from the normal forms. In other instances the crowns are of a conical form, or the point may present a depressed or otherwise irregular surface. Hunter is mistaken in saying that they only occur amongst the incisors and canines. They are occasionally observed amongst the molar series. Sometimes they resemble irregularly-formed wisdom teeth implanted externally to the normal molars. At other times they are not distinguishable in form from ordinary molars and bicuspids.]

---

(1) Tomes' System of Dental Surgery, p. 210.

This happens only in the adult set of Teeth, and is owing to there not being room in the Jaw for this second set, the Jaw-Bone being formed with the first set of teeth, and never increasing afterwards; so that if the adult set does not pass further back, they must over-lap each other, and give the appearance of a second row.

OF THE USE OF THE TEETH SO FAR AS THEY AFFECT THE VOICE.

The Teeth serve principally for mastication; and that use need not be farther explained.

They serve likewise a secondary, or subordinate purpose; giving strength and clearness to the sound of the voice, as is evident from the alteration produced in speaking, when the Teeth are lost.

This alteration, however, may not depend entirely upon the Teeth, but, in some measure, on the other organs of the voice having been accustomed to them; and therefore when they are gone, those other organs may be put out of their common play, and may not be able to adapt themselves so well to this new instrument. Yet I believe that habit in this case has no great effect; for those people seldom or never do get the better of the defect; and young children who are shedding their Teeth, and are, perhaps, without any Fore Teeth for half a year or more, always have that defect in their voice, till the new Teeth come; and as these grow, the voice becomes clear again.

This use seems to be entirely in the Fore Teeth; for the loss of one of these makes a great alteration, and the loss of two or three Grinders seems to have no sensible effect. As an argument for the use of the teeth in modifying the sound of the voice, we may observe, that the Fore Teeth come at a time when the child begins to articulate sounds, and at that time they are so loose in the gums, that they can be of very little service in mastication.

Every defect in speech, arising from this defect in the organ, is generally attended with what we call a lisp. People who have lost all their Teeth, and most old people for that reason, lose, in a great measure, their voice. This arises partly from the loss of the Fore Teeth, but principally from the loss of all the Teeth, and of the Alveolar Processes of both Jaws, by which means the mouth becomes too small for the tongue, and the lips and cheeks become flaccid; insomuch that the nicer movements of these parts, in the articulation of sounds, are obstructed; and thence the words and syllables are indistinctly pronounced and slurred, or run into one another.

UNDER WHAT CLASS DO THE HUMAN TEETH COME.

Natural historians have been at great pains to prove from the Teeth, that man is not a carnivorous animal; but in this, as in many other things, they have not been accurate in their definitions; nor have they determined what a carnivorous animal is.

If they mean an animal that catches and kills his prey with his Teeth, and eats that flesh of his prey just as it is killed, they are in the right; man is not in this sense a carnivorous animal, and, therefore, he has not Teeth like those of a Lion; and this, I presume, is what they mean.

But if their meaning were that the Human Teeth are not fitted for eating meat that has been catched, killed, and dressed by art, in all the various ways that the superiority of the human mind can invent, they are in the wrong. Indeed, from this confined way of thinking it would be hard to say what the human Teeth are fitted for; because, by the same reasoning, man is not a graminivorous animal, as his Teeth are not fitted for pulling vegetable food, &c. They are not made like those of cows or horses, for example.

The light in which we ought to view this subject is, that man is a more perfect or complicated animal than any other; and is not made like others, to come at his food by his Teeth, but by his hands, directed by his superior ingenuity; the Teeth being given only for the purpose of chewing the food, in order to its more easy digestion: and they, as well as his other organs of digestion, are fitted for the conversion of both animal and vegetable substances into blood; and thence he is able to live in a much greater variety of circumstances than any other animal, and has more opportunities of exercising the faculties of his mind. He ought, therefore, to be considered as a compound, fitted equally to live upon flesh and upon vegetables. (*p*)

OF THE DISEASES OF THE TEETH.

The Teeth are subject to diseases as well as other parts of the body. Whatever the disorder is that affects them, it is generally attended with pain; and from this indeed we commonly first know that they are affected.

Pain in the Teeth proceeds, I believe, in a great measure, from the air coming into contact with the nerve in the cavity of the Tooth; for we seldom see people affected with the Tooth-ach, but when the cavity is exposed to the air.

It is not easy to say by what means the cavity comes to be exposed.

The most common disease to which the Teeth are subject, begins with a small, dark coloured speck, generally on the side of the Tooth where it is not exposed to pressure; from what cause this arises is hitherto unknown. The substance of the Tooth thus discoloured, gradually decays, and an opening is made into the cavity. As soon as the air is thereby admitted, a considerable degree of pain arises, which is probably owing to the admission of the air, as it may be prevented by filling the cavity with lead, wax, &c. This pain is not always present; the food, and other substances, perhaps fill up the hole occasionally, and prevent the access of the air, and of consequence the pain, during the time they remain in it. When

---

(*p*) [Professor Owen has observed that man's teeth seem originally formed to " eat of the fruit of the garden."]

an opening is made into the cavity of the Tooth, the inside begins to decay, the cavity becomes larger, the breath at the same time often acquires a putrid *Fœtor*, the bone continues to decay till it is no longer able to support the pressure of the opposite Tooth, it breaks and lays the cavity open. We have not as yet found any means of preventing this disease, or of curing it; all that can be done, is to fill the hole with lead, which prevents the pain, and retards the decay; but after the Tooth is broken, this is not practicable; and for that reason it is then best to extract it. (*q*)

It would be best of all to attempt the extraction of a Tooth by drawing it in the direction of its axis: but that not being practicable by the instruments at present in use, which pull laterally, it is the next best to draw a Tooth to that side where the Alveolar Process is weakest; which is the inside, in the two last grinders on each side of the Lower-Jaw, and the outside in all the others.

It generally happens in drawing a Tooth, that the Alveolar Process is broken, particularly when the Grinders are extracted; but this is attended with no bad consequences, as that part of the Alveolar Process from which the Tooth was extracted always decays.

In drawing a Tooth, the patient complains of a disagreeable jarring noise, which always happens when any thing grates against the bones of the head.

OF CLEANING THE TEETH.

From what was said of the nature and use of the Enamel, it is evident, that whatever is capable of destroying it, must be hurtful;

(*q*) [We do not propose here to give a *résumé* of the modern views as to the pathology and treatment of caries. With regard to the latter it is needless to say that the resources of dental surgery have been widely extended since Hunter's time. In extracting, the use of the forceps has superseded that of the key.]

therefore all acids, gritty powders, and injudicious methods of scaling the Teeth are prejudicial: but simply scaling the Teeth, that is clearing them of the stony concretions which frequently collect about their necks, while nothing is scraped off but that adventitious substance, is proper and useful. (*r*) If not removed by art, the quantity of the stony matter is apt to increase, and to affect the gum. This matter first begins to form on the Tooth near to the Gum; but not in the very angle, because the motion of the Gum commonly prevents the accumulation of it at this part. I have seen it cover not only the whole Tooth, but a great part of the Gum: in this case there is always an accumulation of a very putrid matter, frequently considerable tenderness and ulceration of the Gum, and scaling becomes absolutely necessary.*

---

* The animal fluids, when out of the course of the general circulation, especially when they stagnate in cavities, are apt to deposit an absorbent earth, and form concretions. This earth is sometimes contained in the fluids; and is only deposited; as in the formation of the stone in the urinary passages : in some cases, perhaps, the fluids undergo a change, by which the earth is first formed, and afterwards deposited. This deposition takes place particularly in weakened parts, or where the circulation is languid, or where there are few arteries, such as about joints and tendons ; as if it were intended to strengthen these parts, if they should at any time give way ; for if an artery, for instance, is overcome by the action of the heart, and unnaturally dilated, its coats have commonly these concretions formed everywhere in their interstices. The same

---

(*r*) [Tartar consists of matters precipitated from the saliva, and oral and pulmonary mucus. Epithelial scales, and occasionally infusorial animalcules, contribute to its formation. Simon says : "Tartar on the human teeth consists of earthy phosphates, epithelium scales, a little ptyalin, and fat, and when examined under the microscope, there are seen abundance of pavement epithelium and mucus corpuscles. And in addition to these, numerous long acicular bodies and infusoria of the genera vibrio and monas.] (1)

---

(1) *Vide* Tomes op. cit., p. 536. Simon's Annual Chemistry, translated by Day.

OF TRANSPLANTING THE TEETH.

From considering the almost constant variety of the size and shape of the same class of Teeth in different people, it would appear almost impossible to find the Tooth of one person that should fit, with any degree of exactness, the socket of another; and this observation is supported, and indeed would seem to be proved by observing the Teeth in skeletons. Yet we can actually transplant a Tooth from one person to another, without great difficulty, nature assisting the operation, if it is done in such a way that she can assist; and the only way in which nature can assist, with respect either to size or shape, is by having the fang of the transplanted Tooth rather smaller than the socket. The socket, in this case grows to the Tooth. If the fang is too large, it is impossible indeed to insert it at all in that state; however, if the fang should be originally too large, it may be made less; and this seems to answer the purpose as well.

The success of this operation is founded on a disposition in all living substances, to unite when brought into contact with one another; although they are of a different structure; and even although the circulation is only carried on in one of them.

---

thing happens also in the coats of incysted tumours, which are constantly distended; in cases of distentions of the Tunica Vaginalis Testis, &c. It is also apt to take place in parts which have lost their natural functions; as in the coats of the eye in cases of blindness, and in diseased lymphatic glands, &c., and where the living power is diminished in the system, as in the arteries, membranes, &c., of old people; and in some particular habits, as in those who are affected by the gout.

The same sort of deposition takes place likewise where there is any substance with such properties as render it a fit *Basis* for crystallization; as when extraneous bodies are lodged in the bladder: whence such bodies are so often found to form the *Nucleus* of a stone. The same thing happens in the bowels of many animals; whence the *Nucleus* of intestinal concretions, or bezoars, is commonly a nail, or some indigestible substance which had been swallowed. The crust, which collects upon the Teeth, seems to be a crystallization of the same nature.

This disposition is not so considerable in the more perfect or complex animals, such as quadrupeds, as it is in the more simple or imperfect; nor in old animals, as in young: for the living principle in young animals, and those of simple construction, is not so much confined to, or derived from one part of the body; so that it continues longer in a part separated from their bodies, and even would appear to be generated in it for some time; while a part, separated from an older, or more perfect animal, dies sooner, and would appear to have its life entirely dependent on the body from which it was taken.

Taking off the young spur of a cock, and fixing it to his comb, is an old and well known experiment.

I have also frequently taken out the *Testis* of a cock and replaced it in his belly, where it has adhered, and has been nourished; nay, I have put the *Testis* of a cock into the belly of a hen with the same effect.

In like manner a fresh Tooth, when transplanted from one socket to another, becomes to all appearance a part of that body to which it is now attached, as much as it was of the one from which it was taken; while a Tooth which has been extracted for some time, so as to lose the whole of its life, will never become firm or fixed: the sockets will also in this case acquire the disposition to fill up, which they do not in the case of the insertion of a fresh Tooth. (*s*)

(*s*) [The practice of transplanting teeth from one person to another originated, I believe, with Hunter, under whose superintendence it was frequently performed. Had the results of all these cases been known to him, it is probable that this recommendation would not have been written; there is not, I believe, a single instance of its perfect success, and there are many in which it has been followed by even fatal results. Fox, in his excellent practical work on the teeth, strongly reprobates this practice, and has probably prevented much pain and disease by exposing its continual failure, its occasional injurious results, and the want of correct feeling which seems to be necessarily involved in its performance. The tooth figured by Fox as having been the subject of this operation is now in the collection of Guy's Hospital; its root is deeply eroded by absorption.] T. BELL.

These appearances shew that the living principle exists in the several parts of the body, independent of the influence of the brain, or circulation, and that it subsists by these or is indebted to them for its continuance; and in proportion as animals have less of brain and circulation, the living power has less dependence on them, and becomes a more active principle in itself; and in many animals there is no brain nor circulation, so that this power is capable of being continued equally by all the parts themselves, such animals being nearly similar in this respect to vegetables.

PART THE SECOND.

A PRACTICAL TREATISE ON THE DISEASES OF THE TEETH.

INTRODUCTION.

The importance of the Teeth is such, that they deserve our utmost attention, as well with respect to the preservation of them, when in an healthy state, as to the methods of curing them, when diseased. They require this attention, not only for the preservation of themselves, as instruments useful to the body, but also on account of other parts with which they are connected; for diseases in the Teeth are apt to produce diseases in the neighbouring parts, frequently of very serious consequences; as will evidently appear in the following Treatise.

One might at first imagine, that the diseases of the Teeth must be very simple, and like those which take place every where else in the bony parts of our body; but experience shows the contrary. The Teeth being singular in their structure, and some other circumstances, have diseases peculiar to themselves. These diseases, considered abstractedly, are indeed very simple; but by the relations which the Teeth bear to the body in general, and to the parts with which they are immediately connected, they become extremely complicated. The diseases which may arise in consequence of those of the Teeth, are various; such as Abcesses, Carious Bones, &c.; many of which although proceeding originally from the Teeth, are more the object of the Surgeon than of the Dentist, who will find himself as much at a loss in such cases, as if the Abcess or Carious Bone were in the leg, or any other distant part. All the diseases of the teeth which are common to them with the other parts of the body, should be put under the management of the Physician or Surgeon; but those which are peculiar to the teeth and their connections, belong properly to the Dentist.

It is not my present purpose to enumerate every disease capable of producing such symptoms as may lead us to suspect the Teeth; for the jaws may be affected by almost every kind of disorder. I shall therefore confine myself to the diseases of the Teeth, Gums, and Alveolar Processes; which parts having a peculiar connection, their diseases fall properly within the province of the Dentist. I shall also purposely avoid entering into common Surgery; not to lead the Dentist beyond his depth and to matters of which it is to be supposed he has not acquired a competent knowledge.

In order that the reader may perfectly understand what follows it will be necessary for him previously to consider and comprehend the anatomy and uses of every part of a Tooth, as explained in my 'Natural History of the Human Teeth,' to which I shall be obliged frequently to refer. Without such previous study, the Dentist will often be at a loss to account for many of the diseases and symptoms mentioned here, and will retain many vulgar errors imbibed by conversing with ignorant people or by reading books in which the anatomy and physiology of the Teeth are treated without a sufficient knowledge of the subject.

Whichever of the connected parts be originally diseased, the Teeth are commonly the greatest sufferers. None of those parts can be distempered, without communicating to the Teeth such morbid effects, as tend to the destruction of them.

CHAPTER I.

OF THE DISEASES OF THE TEETH AND THE CONSEQUENCE OF THEM.

§ 1. *The decay of the teeth arising from rottenness.*

The most common disease to which the Teeth are exposed, is such a decay as would appear to deserve the name of mortification. But there was something more; for the simple death of the part would produce but little effect, as we find that Teeth are not subject to putrefaction after death; and therefore I am apt to suspect, that, during life, there is some operation going on, which produces a change in the diseased part. (*a*) It almost always begins externally in a small part of the body of the tooth, and commonly appears at first as an opaque white spot. This is owing to the enamel's losing its regular and crystalized texture, and being reduced to a state of powder, from the attraction of cohesion being destroyed; which pro-

---

(*a*) [Hunter's conjecture that other changes take place in the diseased part besides decomposition is perfectly correct. If a section is taken of a tooth in which caries has just commenced, and examined by the microscope, the dentinal tubes corresponding to the seat of the disease will be found more or less altered down to the pulp cavity. This part of the dentine is more transparent than elsewhere, owing to the consolidation of the tubes by the deposition of calcareous matter. In some instances, new dentine is also added opposite to the openings of the dentinal tubes on the pulp surface, an evident attempt on the part of nature to prevent the disease from penetrating to the pulp cavity. This formation of *secondary* dentine, as it is sometimes termed, seems to depend upon the rapidity with which the disease advances. If the caries proceeds very rapidly, there does not appear to be such a condition of the parts as to allow of its formation; but if its progress is slow, then new dentine will be added, the irritation caused to the pulp being no more than excites it to increased activity.]

duces similar effects to those of powdered crystal. When this has crumbled away, the bony part of the Tooth is exposed; and when the disease has attacked this part, it generally appears like a dark brown spec. Sometimes however, there is no change of colour, and therefore the disease is not observable, till it has made a considerable hole in the Tooth. The dead part is generally at first round, but not always; its particular figure depending more on the place where it begins, than on any other circumstance It is often observed in the hollow parts of the grinding surface of the *Molares*, and there looks like a crack filled with a very black substance. In the incisors, the disease usually begins pretty near the neck of the Tooth, and the scooping process goes on enlarging the cavity, commonly across the same part of the Tooth, which almost divides it into two. When such a diseased Tooth gives way, the mischief is occasioned by its body breaking off.

When it attacks the bony part it appears first to destroy the earth, for the bone becomes softer and softer, and is at last so soft on the exterior exposed surface, that it can be picked away with a pin, and when allowed to dry, it cracks like dried clay.

It begins sometimes in the inside of the Tooth, although but rarely. In this case the Tooth becomes of a shining black, from the dark colour being seen through the remaining shell of the Tooth, and no hole is found leading into the cavity. (*b*)

This blackness is seldom more than a portion of the bony part

---

(*b*) [Mr. Bell, who regards inflammation as the true proximate cause of dental caries, asserts that the situation in which the disease commences is invariably under the enamel upon the surface of the dentine. All recent authorities are, however, agreed that caries commences on the surface of the enamel, and gradually makes its way to the pulp cavity. Those who have maintained the contrary opinion have been misled by the spreading of the disease under the enamel, while the surface of the enamel in some rare instances appears to be almost intact. A careful examination of such teeth will, however, show the presence of cracks or fissures in the enamel through which the fluids of the mouth have found their way to the surface of the dentine.]

decayed or mortified. However, it often happens, that the remaining part of the Tooth becomes simply dead; in which state it is capable of taking on a dye. As it is generally on the external surface one might expect no great mischief would ensue; but the tendency to mortification goes deeper and deeper, till at last it arrives at the cavity of the Tooth, and the mortification follows. Mortification is common to every part of the body: but in most other parts, this tendency is owing in a great measure to the constitution, which being corrected, that disposition ceases; but here it is local, and as such it would appear that we have no power of resisting it. When gone thus far, the decay makes a quicker progress similar to those cases where the decay begins in the cavity; for then this disposition is given to the whole cavity of the Tooth, which, being a much larger surface than what the disease had before to act upon, the increase of the decay seems to be in the same proportion: at last it scoops out its inner substance, till almost nothing is left but a thin shell, which generally, being broken by mastication, a smaller or larger opening is made, and the whole cavity becomes at length exposed.

The canal in the Fang of the Tooth is more slowly affected: the scooping process appears to stop there, for we seldom know a Fang become very hollow to its point, when in the form of a stump: and it sometimes appears Sound, even when the body of the Tooth is almost destroyed; hence I conclude, that the Fang of the Tooth has greater living powers than the Body, by which the process of the disease is retarded, and this part appears at last only to lose its living principle, and not take on the mortifying process above described; for which reason it remains simply a dead Fang; however it does not remain perfectly at rest.

This is the stage in which it is called a Stump. It begins now to lose its sensibility, and is seldom afterwards the cause of pain. (c)

(c) [Stumps which remain in the mouth, and in which decay becomes

Thus in appearance, it will remain sometimes for many years but there will be more or less of a change going on; Nature will be attempting to make up the deficiency, by endeavouring to increase the Stump; for in many cases we find the Stumps thickened and lengthened at their terminations, or small ends; but it is a process she is not equal to, therefore no advantages accrue from it. When she either fails in this process, or is in such a state as not to attempt it, then by this condition of the Tooth a stimulus is given to the alveolar processes, which produces a filling up of the socket from the bottom, whereby the Stumps are gradually protruded. But although they are pushed out at the bottom, they seldom or never project farther beyond the gum than at first; and that part of the Tooth which projects, seems to decay in proportion to its projection.

---

arrested, have a semi-transparent waxy look, arising from the consolidation of the dentinal tubes previously referred to as one of the effects of caries, and from the formation of new dentine by which the canal of the fang has become partially closed up. These stumps are not entirely dead, but retain a low amount of vitality from the supply of the nutrient fluid received from the periosteum, and investing layer of cementum. They often remain for years without producing any annoyance, until inflammation is set up in the periosteum. Should caries arise in a stump, it generally attacks the centre, which, becoming hollow and worn away, forms a kind of cup where particles of food and the fluids of the mouth accumulate, and by their direct action or their decomposition, give rise to destruction of the dentine. This destructive process proceeds towards the exterior of the fang, and downwards towards the apex, until at length the stump is reduced to a conical hollow shell of osseous matter. While this is going on, one of two things happens, either inflammation arises in the periosteum producing great pain, and terminating in a purulent discharge, which necessitates the extraction of the stump, or the stump, gradually losing what slight amount of vitality it possessed, becomes a foreign body, and nature, by a double process of absorption, the one taking place around the apex of the stump, the other in the surrounding tissues, ultimately expels it from the mouth. Stumps are often the seat of exostosis; when, however, they are about to be expelled from the mouth, this is absorbed, leaving a ragged projecting extremity, which is sometimes mistaken for a new growth.]

Besides this decay at the external end of the Stump, there is an absorption of the Fang at the bottom, which is known by the following observation: the end of the Stump, which was in the gum or jaw, becomes irregularly blunted, and often rough, and has not the appearance of the end of the Fang of a Sound Tooth.

Such Stumps are in general easily extracted, being attached often to little more than the gum, and that sometimes loosely.

Although the disease appears to be chiefly in the Tooth itself, and but little to depend on external causes, yet in many cases the part which is already rotten, seems to have some influence upon that which remains: for if the rotten part be perfectly removed before it has arrived at the canal of the Tooth, a stop is sometimes put to the farther progress of the decay, at least for a time. (*d*)

However this is not constantly so; it is oftener the contrary; but it is expedient in most cases to make this trial, as it is always right to keep a Tooth clean, and free from specks.

This decay of the Teeth does not seem to be so entirely the effect of accident as might be imagined; for it sometimes takes place in them by pairs, in which case we may suppose it owing to an original cause coming into action at its stated time; the corresponding Teeth being in pairs, with respect to the disease, as well as to situation, shape, &c.

This opinion is somewhat strengthened by the fore Teeth in the lower Jaw not being so subject to decay as those in the upper, although equally liable to all accidents arising from external influence, which could produce the disease in general.

(*d*) [Diseased teeth are not the direct cause of caries in others, but inasmuch as they allow the food and the fluids of the mouth to accumulate and to undergo decomposition, they favour its development and progress.]

The fore Teeth in the lower Jaw appear to be less subject to this disease than any of the others; the fore Teeth in the upper Jaw, and the grinders in both, are of course more frequently affected.

This disease and its consequences seem to be peculiar to youth and middle age; the shedding Teeth are as subject to it, if not more so, than those intended to last through life; and we seldom or ever see any person, whose Teeth begin to rot after the age of fifty years.

This might be supposed to arise from the disproportion that the number of Teeth, after fifty, bear to them before it; but the number of diseased teeth after fifty do not bear the same proportion. (e)

(e) [Numerous theories and hypotheses have from time to time been put forth as to the cause of dental caries. These causes may be arranged under the two heads of *predisposing* causes and of *exciting* or *proximate* causes. Among the predisposing causes may be enumerated original imperfections in the tissues of the teeth, such as cracks or imperfections in the enamel, and white opaque imperfectly-formed spots. The honey-combed depressions which are so often seen in the enamel, and also the transverse depressions which surround the crown of the tooth like so many rings, are sure indications of an imperfectly formed dentine. Microscopic sections of such teeth show the existence of open spaces in the dentine, where calcification has not taken place. These spaces are black and opaque when filled with air, or clear and apparently void, when the tooth is saturated with a transparent fluid, such as oil, water, or canada balsam. Crowding irregularity, and want of care in cleaning the teeth, must also be classed amongst the predisposing causes.

Another and important predisposing cause is the occurrence of certain constitutional disorders which exhaust the powers of the body, leaving the patient in a state of extreme debility, and with a long period of convalescence before he is restored to health. It is well known that the eruptive fevers of childhood are often followed by the loss of a portion of the jaw bones, together with the included germs of the permanent teeth. (1) In a similar manner the more severe fevers of adult life

---

(1) See Mr. **Salter's cases and remarks.** Guy's Hospital Reports, Vol. VI. Third Series.

This disease has not hitherto been accounted for; if it had been always on the inside of the cavity it might have been

are apt to be followed by rapid and general caries of the permanent teeth. Cholera, and all diseases accompanied with extreme prostration of the vital powers, and cases attended with great loss of blood are liable to be followed by a rapid decay of the teeth.

Certain climates or localities would appear to predispose the inhabitants to diseases of the teeth. Thus, what has been termed expulsive gingivitis, prevails in the district of Lower Austria, extending from the south of the plains of Vienna as far as the province of Styria. (1) These effects of climate are exerted on the membranes rather than upon the teeth themselves.

The exciting causes include an acid condition of the fluids of the mouth, certain derangements of the digestive organs, which are accompanied by acid eructations, the use of acids such as vinegar and lemon juice as articles of food, and the employment of the stronger acids when administered as medicines.

Various theories have been advanced from time to time in explanation of dental caries, but they may all be referred to one of the three following:—the inflammatory, the chemical, or the chemico-vital.

Fox says, "The proximate cause of caries appears to be an inflammation in the bone of the crown of the tooth, which, on account of its peculiar structure, terminates in mortification." This writer, however, would appear to have confounded caries, or partial death of the tooth, with necrosis, or total death of the tooth, for he remarks, "If a sound tooth, that has been recently extracted, be broken, the membrane will be found to be firmly attached to the bone of the tooth, forming the inner cavity. But when this membrane becomes inflamed, it separates from the bone, and the death is the consequence."

That this is the proximate cause of caries, appears to be highly probable, by remarking that caries of other bones is caused by a separation of those membranes which cover them, and which are attached to them." (2)

Mr. Bell, who proposes to substitute the term GANGRENE of the tooth for caries, and defines it as "mortification of any part of the tooth, producing gradual decomposition," also asserts that inflammation is its proximate cause: "Still, however, the true proxi-

---

(1) See Lectures on Diseases of the Dental Periosteum, by R. T. Hulme, p. 54. London, 1862.

(2) The Natural History and Diseases of the Human Teeth, by Joseph Fox, Pt. 2, p. 12. Second Edition. London, 1814.

supposed to have been owing to the deficiency of nourishment from some fault in the vascular system; but as it begins

mate cause of dental gangrene is inflammation; and the following appears to me to be the manner in which it takes place; when from cold or any other cause, a tooth becomes inflamed, the part which suffers the most severely, is unable, from its possessing comparatively but a small degree of vital power, to recover from the effects of inflammation; and mortification of that part is the consequence." (1)

Oudet refers the production of caries to original imperfections in the structure of the dentine, its progress is from within outwards, and it is often hereditary. "Caries proceeds from within outwards. Injured in its vitality, by means which escape our notice, but which must sometimes affect the pulp, whose delicate tissues do not enable it to resist the external agents with which the teeth come in contact, the ivory (dentine) becomes the seat of alterations, which affect its colour, and the cohesion of its molecules. It becomes of a yellow or brown colour, and softens in consequence of the changes which have taken place in it. Greatly altered in its texture and its composition, it acquires new chemical properties, and becomes an agent of destruction to the surrounding tissues. The change in the ivory soon extends to the enamel, gradually invading it as it spreads to the surface of the tooth. It excavates a cavity which enlarges with the progress of the disease, reduces the enamel to its superficial layers, until this substance, deprived of support, breaks away and discovers the caries." (2)

In order to account for these internal changes in the structure of the dentine, Oudet gratuitously assumes that an acid is generated in the imperfectly formed tissue which becomes the agent of its destruction.

Mr. Robertson may be taken as the exponent of the chemical theory of caries, and has employed many ingenious arguments to prove its correctness; but which are summed up in the following passages.

"From the review which we have just taken of decay in the different classes of teeth, it will be perceived that, in regard to situation, it takes place on the surfaces of the teeth, in excavations found in them and the projecting gum, in cavities, indentations, and irregularities on the external substance of the tooth itself, and it occurs at their sides, in their

---

(1) The Anatomy, Physiology, and Diseases of the Teeth, by Thomas Bell, F.R.S. Second Edition. P. 126. London, 1835.

(2) Recherches sur les Dents et sur leurs Maladies. Par J. E. Oudet, p. 75, Paris, 1862.

most commonly externally, in a part where the teeth in their most sound state receive little or no nourishment, we cannot refer to that cause.

necks, and in the spaces produced by their formation and relative position ; in regard to frequency, that it is in proportion to the depth of the superficial depressions, and the degree and nature of the lateral projections and interstices.

" This being the case, and decay never being found to commence upon the plain and smooth surface of the tooth, it cannot for a moment be doubted that the predisposition to caries depends upon the external configuration or conformation of the teeth. It must be equally evident from the partial nature of the disease, and from the insufficiency of all general causes, as before pointed out, to explain its origin, that the exciting cause of caries must be one whose operation is partial, and having a peculiar action upon those parts of the teeth, which are by their action predisposed to decay. The only cause of the partial operation and the particular situations of decay, is the corrosive or chemical action of the solid particles of the food which have been retained and undergone a process of putrefaction or fermentation in the several parts of the teeth best adapted for their lodgment." (1)

Harris entertains similar opinions upon the nature of caries. According to this writer : " Caries of the teeth is the result of the action of chemical agents, and not that of any operation of the animal economy, and it consists simply in the decomposition of the calcareous molecules of the organs. The fluids of the mouth, especially the mucous, when in a vitiated condition, contain an acid, namely, the septic (nitrous), which has a strong affinity for the earthy ingredients of these organs ; and it is by the action of the former upon the latter that the affection is produced."(2)

Mr. Tomes, in his Lectures on Dental Surgery (p. 199), has promulgated what has been termed the chemico-vital theory of caries. " If," he says, " the views I am about to explain are sound, caries may be defined to be *death and subsequent progressive decomposition of a part or the whole of a tooth.*

"I believe that the dentine from abnormal action loses its vitality, and with the loss of vitality the power of resisting chemical action, and that consequently the dead part is, under favouring circumstances, de-

---

(1) A Practical Treatise on the Human Teeth, by William Robertson, pp. 36. London, 1846.
(2) See the American edition of Fox's work, with notes by Chapin A. Harris, M.D., pp. 161. Also Harris's Dental Surgery.

It does not arise from any external injury, or from menstrua, which have a power of dissolving part of a tooth; for anything of that kind could not act so partially; and we can observe in those Teeth where the disease has not gone deep, that from the black speck externally there is a gradual decay

---

composed by the fluids of the mouth. That there must be a concurrence of dead dental tissue, and of a condition of the oral fluids capable of decomposing the dead part before the phenomena of caries can be developed.

"Further, I conceive that the causes producing the abnormal action may have been applied locally to the tooth itself, or may have had a constitutional origin, and therefore have acted through the nerves or the circulating fluids."

The latter theory appears to afford the most correct explanation of the nature of dental caries. At the same time there is scarcely sufficient proof that the part is actually dead before caries can commence, and it would perhaps be more correct to say that the vitality of the part is so far diminished or altered as to render it unable to resist the chemical action of external agents. One point of some importance seems to have been overlooked by most writers on caries. Great attention has been paid to the condition of the saliva, but the state of the pulp and of the nutrient fluid which it conveys into the dentine has been altogether omitted. It is well known that the serum of the blood is alkaline, and Professor Wurtz found, upon analysing the pulp of the tooth, that this also was strongly alkaline; hence it is only reasonable to conclude that the fluid pervading the dentine is the same. The origin of caries may therefore depend upon the relative state of the alkaline and acid conditions of the pulp and of the saliva, not only may it arise from excessive acidity of the saliva, but it may also depend upon a want of alkali in the pulp and fluids of the tooth.

While nearly every recent authority attributes caries to the action of an acid, it is still a question what is the nature of this acid, and how it is generated. Harris mentions nitrous acid, but I know not upon what authority, nor am I aware of any proof that decomposing food gives rise to this acid. In a paper read at one of the meetings of the Odontological Society, Mr. Bridgeman endeavoured to show that the destructive action was produced by lactic acid, founding his views upon the statement that moist animal membrane in a slightly decaying condition often acts energetically in developing lactic acid. (1)]

---

(1) See British Journal of Dental Science. Vol. 4.

or alteration leading to the cavity, and becoming fainter and fainter. We may therefore reasonably suppose, that it is a disease arising originally in the Tooth itself; because when once the shell of the Tooth has given way to the cavity, the cavity itself soon becomes diseased in the same way. That the disease spreads thus rapidly over the cavity, as soon as the Tooth has given way, does not depend simply on the exposure; for if a sound Tooth be broken by accident, so as to expose the cavity, no such quick decay ensues: however, sometimes we find in those cases, that exposure of the cavity will produce a decay, and even pain, similar to an original disease; and in the diseased Tooth we find that the exposure has a considerable effect in hastening the progress of the disease; for if the Tooth be stopped so as to prevent its exposure to external injury, its cavity will not nearly so soon become diseased. Exposure therefore seems at least to assist the decay.

How far a rotten Tooth has the power of contaminating those next to it, I believe is not yet completely ascertained; some cases seem to favour this idea, and many to contradict it. We frequently see two Teeth rotten in places exactly opposite to each other, and as one of them began first to decay it gives a suspicion that the last diseased was infected by that which received the first morbid impression.

On the contrary, we often see one diseased, whilst another Tooth, in contact with the decayed part, remains perfectly sound.

SYMPTOMS OF INFLAMMATION.

Few or no symptoms are produced by this disease, besides the above appearances, till the cavity of the Tooth is exposed; however it often happens, that a tenderness, or a soreness upon touch, or other external influences, takes place long before; but when the cavity is exposed, then pain and other symptoms often begin, which are generally very considerable: however, the exposure of the cavity of a tooth does not in all cases give pain. Some Teeth shall moulder wholly away, without ever having any sensation. (*f*)

In many cases, there will be very acute pain upon the cavity being exposed, which will subside, and recur again, without producing any other effect; but it more frequently happens, that this pain is the first symptom of inflammation, and is in most cases very considerable; more so than that arising from such an inflammation in other places. The surrounding parts sympathize commonly to a considerable extent, viz. the Gums, Jaw bones, and integuments covering them; they inflame and swell so much as to affect the whole of that side of the face, where the affected Tooth is situated. The mouth can hardly be opened; the glands of that side of the neck often swell; there is an increase of the saliva, and the eye is almost closed; the Tooth not giving way to the swelling of the soft parts within it: and for this reason the local effects of the Inflammation cannot be so visible as in the soft parts.

This inflammation of the Tooth often lasts a considerable time, and then gradually subsides. We may suppose, according to the general law of inflammation, that it is first of the adhesive kind, and accordingly we sometimes find the Teeth swelled at their ends, which is a character of the adhesive stage of inflammation; and sometimes two fangs are grown together. That we seldom find adhesions between the Teeth and surrounding parts may be reasonably imputed to their less aptitude for such connections. The suppurative inflammation succeeds; but as a Tooth has not that power of suppuration which leads

---

(*f*) [The occurrence of pain would appear to depend very much upon the rapidity with which the disease advances. When caries runs a very rapid course, pain speedily ensues, but where its progress is slow the crown of the tooth may be destroyed without any violent attack of pain. This arises from time being given for the pulp to form new dentine which gradually fills up the cavity of the tooth as the caries advances towards the centre.]

to granulations, so as to be buried, covered up, and made part of ourselves, as happens to other bones, (which would destroy any use of a Tooth) the inflammation wears out, or rather the parts not being susceptible of this irritation, beyond a certain time, the inflammation gradually goes off, and leaves the Tooth in its original diseased state. No permanent cure therefore can possibly be effected by such inflammations, but the parts being left in the same state as before, they are still subject to repetitions of inflammation, till some change takes place, preventing future attacks, which I believe is generally, if not always, effected by the destruction of those parts which are the seat of it, *viz.* the soft parts within the Tooth. (*g*)

Nature seems in some measure, to have considered the Teeth as aliens, only giving them nourishment while sound and fit for service, but not allowing them when diseased the common benefits of that society in which they are placed. They cannot exfoliate, as no operations go on in them except growth; therefore, if any part is dead, the living has not the power of throwing it off, and forming an external surface capable of supporting itself, like the other parts of the body: indeed, if they had

---

(*g*) [In the previous paragraph the Author has been speaking of caries, but in the present he evidently refers to inflammation of the periosteum, for it is not until this membrane has become affected, that the symptoms which he has described take place. Hunter, in fact, confounds together the inflammation of the pulp and of the periosteum of the tooth. To say that the surrounding parts have "less aptitude for such connexion" is only another mode of expressing the fact that union between a tooth and its socket is extremely rare, if, indeed, it ever takes place. The non-occurrence of this union appears to the writer to show that the periosteum is not common to the socket and the fang of the tooth as asserted by Hunter, nor is it a reflection of that which lines the socket as stated by Bell, but is formed by a prolongation of the outermost membrane of the tooth germ; the periosteum of the alveolus being distinct and consisisting of a very delicate membrane resembling that which lines the medullary cavity of the long bones.]

such a power, no good purpose could be answered by it; for a piece of Tooth, simply dead, is almost as useful as if the whole was living: which may be observed every day. (*h*)

The pain, however, appears to take its rise from the Tooth as a centre. That it should be more severe than what is generally produced by similar inflammations in other parts of the body, may, perhaps, be accounted for, when we consider, that these parts do not readily yield; as likewise the case in whitloes.

It sometimes happens, that the mind is not directed to the real seat of the disease, the sensation of pain not seeming to be in the diseased Tooth, but in some neighbouring Tooth which is perfectly sound. This has often misled operators, and the sympathising Tooth has fallen a sacrifice to their ignorance.

In all cases of diseased Teeth, the pain is brought on by circumstances unconnected with the disease; as for instance Cold, wherefore they are more troublesome commonly in winter than in summer. Extraneous matter entering the cavity, and touching the nerve and vessels, will also bring on the pain.

This pain is frequently observed to be periodical; sometimes there being a perfect intermission, sometimes only an abatement of it. The paroxysm comes on once in twenty four hours; and, for the most part towards the evening. The bark has therefore been tried; but that failing, the disorder has been suspected to be of the rheumatic kind, and treated accordingly

---

(*h*) [Although the teeth cannot reproduce a portion of dentine which has once been destroyed, nature has not left them without some means of resisting the effects of disease. The filling up of the pulp cavity in the manner previously mentioned by the formation of new dentine is evidently for the purpose of preserving the tooth. It is owing to this that teeth which could not bear stopping when first seen, may be brought into such a condition as to render the operation practicable. Many cases present themselves where, by clearing out the decayed cavity, removing all extraneous matter, and inserting a temporary stopping, or by inserting cotton moistened with spirit, and renewing the application daily for a month or six weeks, a permanent stopping can be successfully applied.]

with no better success. At length, after a more particular examination of the Teeth, one of them has been suspected to be unsound; and, being extracted has put an end to the disorder. This shows how injudicious it is to give medicine in such cases, while the true state of the Tooth is unknown.

This disease is often the cause of bad breath more so than any other disease of those parts; especially when it has exposed the cavity of the Tooth. This most probably arises from the rotten part of the Tooth, and the juices of the mouth, and food, all stagnating in this hollow part, which is warm, and hastens putrefaction in them.

I come now to the prevention and cure of this disease.

The first thing to be considered, is the cure of the decaying state of the Tooth, or rather the means of preventing the further progress of the decay; and more especially before it hath reached the cavity, whereby the Tooth may be in some degree preserved; the consequent pain and inflammations, commonly called Tooth-ach, avoided and often the consequent abcesses called Gum Boils. I believe, however, that no such means of absolute prevention are as yet known. The progress of the disease, in some cases, appears to have been retarded, by removing that part which is already decayed; but experience shews, that there is but little dependence upon this practice. I have known cases, where the black spot having been filed off, and scooped entirely out, the decay has stopped for many years. This practice is supposed to prevent at least any effect, that the part already rotten may have upon the sounder parts; however if this is all the good that arises from this practice, I believe in most cases, it might be as well omitted. Even if it were an effectual practice, it could not be an universal one; for it is not always in the power of the operator to remove this decayed part either on account of its situation, or on account of its having made too great a progress, before it is discovered. When it is on the basis of the grinder, or on the posterior side of its neck, it can scarcely be reached. It becomes also impracticable, when the disease is still allowed to go on, and the cavity becomes exposed, so that the patient is now liable to all the consequences already described, and the Tooth is making haste towards a total decay; in such a case, if the decay be not too far advanced, that is, if it be not rendered useless simply as a Tooth, I would advise that it be extracted, then immediately boiled, with a view to make it perfectly clean, and also destroy any life there may be in the Tooth; and then that it be restored to the socket: this will prevent any farther decay of the Tooth, as it is now dead, and not to be acted upon by any disease, but can only suffer chymically or mechanically. (*i*)

This practice, however I would only recommend in grinders, where we have no other resource on account of the number of fangs, as will be more fully explained hereafter. This practice has sometimes been followed with success; and when it does succeed, it answers the same end as the burning the nerve, but with much greater certainty.

If the patient will not submit to have the Tooth drawn, the nerve may be burned: that this may have the desired effect, it must be done to the very point of the fang, which is not always possible. Either of the concentrated acids, such as those of vitriol, nitre, or sea salt, introduced as far into the fang of the Tooth as possible, is capable of destroying its soft parts, which most probably are the seat of pain: a little caustic alkali will produce the same effect. But it is a difficult operation to introduce any of these substances into the root of the fang, till the decay has gone a considerable length, especially, if it be a Tooth of the upper jaw; for it is hardly possible to make fluids pass against their own gravity: in these cases, the common caustic is the best application as it is a solid. The caustic should be intro-

---

(*i*) [It is needless to dwell upon this suggestion further than to remark that such an operation cannot be countenanced in the present day. In the majority of cases it would give rise to severe inflammation and might be followed by serious results.]

duced with a small dossil of lint, but even this will scarcely convey it far enough. If it be the lower jaw, the caustic need only be introduced into the hollow of the Tooth, for by its becoming fluid, by the moisture of the part it will then descend down the cavity of the fang as will also any of the acids; but patients will often not suffer this to be done, till they have endured much pain, and several inflammations.

When there is no other symptom except pain in the Tooth, we have many modes of treatment recommended, which can only be temporary in their effects. These act by derivation, or stimulus applied to some other part of the body. Thus to burn the ear by hot irons, has sometimes been a successful practice, and has relieved the Tooth-ach.

Some stimulating medicine, as spirit of lavender, snuffed up into the nose will often carry off the pain.

When an inflammation takes place in the surrounding parts, it often is assisted by an additional cause, as cold, or fever; when the inflammation hath taken place in a great degree, then it becomes more the object of another consideration; for it may be lessened like any other inflammation arising from similar causes, the pressure of an extraneous body, or exposure of an internal cavity.

If the inflammation be very great, it will be proper to take away some blood. The patient may likewise properly be advised to hold some strong vinus spirit for a considerable time in his mouth. Diluted acids, as vinegar, &c. may likewise be of use applied in the same manner. Likewise, preparations of lead would be advisable; but these might prove dangerous, if they should be accidentally swallowed.

If the skin is affected, poultices, containing some of the above mentioned substances produce relief. The pain, in many cases, being often more than the patient can well bear, warm applications to the part have been recommended such as hot brandy, to divert the mind; also spices, essential oils, &c. which last are, perhaps, the best. A little lint or cotton soaked in laudanum, is often applied with success, and laudanum ought likewise to be taken internally, to procure an interval of some ease. Blisters are of service in most inflammations of these parts, whether they arise from a diseased Tooth or not. They cannot be applied to the part, but they divert the pain, and draw this stimulus to another part; they may be conveniently placed either behind the ear, or in the nape of the neck. These last-mentioned methods can only be considered as temporary means of relief, and such as only affect the inflammation. Therefore the Tooth is still exposed to future attacks of the same disease. (*j*).

STOPPING OF THE TEETH.

If the destruction of the life of the Tooth, either by drawing and restoring it again, or by the actual or potenial cauteries, has not been effected, and only the cure of the inflammation has been attempted, another method of preventing inflammation is to be followed, which is to allow as little stimulus to take place as possible. The cavity of the Tooth not being capable of taking the alarm like most other cavities in the body, and of course not suppurating, as has been already observed, often no more is necessary, either to prevent the inflammation from taking place altogether, or extending farther, than to exclude all extraneous irritating matter ; therefore, the stopping up the cavity becomes, in many cases, the means of preventing future

---

(*j*) [The only justifiable method of applying the actual cautery is by means of the galvanic battery, and even this is seldom resorted to, most practitioners preferring the use of chemical agents. The stronger acids are objectionable on account of the injury they cause to the dentine, and the impossibility of limiting their action. Chloride of zinc, arsenic, and creasote, are the substances most frequently employed.

The remedies mentioned are none of them such as should be used. It must be a very exceptional case in which general blood-letting should be resorted to, and in no instance where the tooth could be extracted. Local bleeding by leeches is often useful when the periosteum is affected. Diluted acids would injure the teeth, and it is difficult to understand how the salts of lead could give relief.]

attacks of the inflammation, and often retards even the progress of the disease, that is, the farther decay of the Tooth, so that many people go on for years thus assisted: but it is a method which must be put in practice early, otherwise it cannot be continued long; for, if the disease has done considerable damage to the inside of the Tooth, so as to have weakened it much, the whole body of the Tooth, most probably, will soon give way in mastication: therefore, under such circumstances, the patient must be cautioned not to make too free with the Tooth in eating. (*k*)

Gold and lead are the metals generally made use of for stopping Teeth. Gold being less pliable, must be used in the leaf; lead is so soft in any form, as to take on any shape by a small force.

Stuffing the hollow Tooth with wax, galbanum, &c. can be of very little service, as it is in most cases impossible to confine these substances, or preserve them from being soon worn away;

(*k*) [Inflammation, in the ordinary acceptation of the term, cannot take place in the dentine or enamel of the tooth, it may do so in the cementum and proceed to suppuration as shown by one specimen in the possession of the writer. Mr. Bell has indeed related a case in which he considered suppuration had occurred in the crown of a molar tooth. No other instance has ever been recorded by any other writer, and as it is stated that the cavity of the abscess communicated with the natural cavity of the tooth, it is more reasonable to conclude the pus had proceeded from the pulp. The remarks which have been previously made upon caries tend to prove that it is essentially a chemical process, and hence it is easily understood that by hermetically closing up a carious cavity the pain arising from the contact of extraneous matter may be relieved, and the further progress of the disorder arrested.

The increased knowledge which we possess of the physiology and pathology of the teeth has taught us that if time is allowed for the deposit of new dentine as mentioned at p. 161, and at the same time the decayed cavity is preserved from the contact of the food and of the fluids of the mouth, teeth may eventually be stopped and preserved for many years, which would formerly have been condemned to extraction.]

however, they have their uses, as it is a practice which the patients themselves can easily put into execution. (*l*)

It often happens from neglect, and much oftener in spite of all the means that can be used, that the Tooth becomes so hollow, as to give way, whereby the passage becomes too large to keep in any of the above-mentioned substances; however in this case it sometimes happens, that a considerable part of the body of the Tooth will still stand, and then a small hole may be drilled through this part, and after the cavity hath been well stopped, a small peg may be put into the hole, so as to keep in the lead, gold, &c. But when this cannot be done, we may consider the broken Tooth as entirely useless, or at least it will soon be so; and it is now open to attacks of inflammation, which the patient

---

(*l*) [Of the various materials in use for the purpose of stopping teeth, gold is still preferred to every other. The gold leaf at present in use is of two distinct qualities, the one is termed adhesive and, after annealing, separate pieces of it are capable of being united into a solid mass by moderate pressure; the other is termed non-adhesive, and separate portions cannot be made to unite; in using the latter for the purpose of stopping, the operator depends upon wedging the layers so closely together that no fluid can penetrate between the folds inserted into the decayed cavity.

The reader is referred to Tomes's System of Dental Surgery, or to Arthur's Treatise on the Use of Adhesive Gold Foil, Philadelphia, 1857, for further information upon this portion of the subject.

Lead is now excluded from use, but when a cheap material is required tin foil is sometimes employed. Other substances that are used in cases where gold cannot be applied consist of the various amalgams, of the so termed osteo-plastic compounds, of Jacob's gutta percha stopping, and some other materials.

Temporary stoppings are very beneficial when a tooth has to be under treatment for some time before filling the cavity, and of these none is perhaps better than gum mastic dissolved in spirits of wine mixed up with a portion of cotton wool sufficiently large, when all the superfluous portion of the mastic solution has been pressed out to fill the cavity. Camphorated spirits of wine applied in the same manner is exceedingly useful where the cavity is so placed that the patient can daily remove the stopping and replace it for himself.]

must either bear, or submit to have the Tooth pulled out. If the first be chosen, and the repeated inflammations submitted to, a cure will be performed in time, by the stump becoming totally dead; but it is better to have it pulled out, and suffer once for all. (*m*)

Upon pulling out these Teeth, we may in general observe a pulpy substance at the root of the fang, so firmly adhering to the fang, as to be pulled out with it. This is in some pretty large, so as to have made a considerable cavity at the bottom of the socket. This substance is the first beginning of the formation of a Gum Boil, as it at times inflames and suppurates. (*n*)

§ 2. *The Decay of the Teeth by Denudation.*

There is another decay of the Teeth much less common than that already described, which has a very singular appearance.

(*m*) [The filling of stumps with gold has been much advocated of late by American dentists, and if the stump is in a healthy condition, and there is no discharge issuing from the cavity, it is far preferable to allowing it to remain open for the lodgment of food and the fluids of the mouth. Much care and judgment is required in selecting these cases; when there has once been inflammation of the periosteum, it is very liable to recur, and certain to do so if the tooth is stopped, when there is a discharge coming from the interior through the canal of the fang.

For a full account of the process of stopping teeth the reader is referred to the various works on Dental Surgery. The volumes of the Dental Cosmos contains many valuable contributions upon recent improvements in stopping.]

(*n*) [The substance to which Hunter refers is not always the first beginning of a gum boil, but may be an adventitious growth consisting of one or more of the histological elements of the periosteum. These growths have been particularly described by M. Magitot.] (1)

---

(1) Memoire sur les Tremeurs de Perioste Dentaire, J. B. Baillière et Fils, Paris, 1860.

A translation [of this Memoir will be found in the Dental Review. Vol. 2. 1860.

See also Lectures on Diseases of the Dental Periosteum in Contributions to Dental Pathology, by R. T. Hulme. H. Baillière. London, 1862.

It is a wasting of the substance of the Tooth very different from the former. In all the instances I have seen, it has begun on the exterior surface of the Tooth, pretty close to the arch of the gum. The first appearance is a want of enamel, whereby the bony part is left exposed, but neither the enamel nor the bony part alter in consistence as in the above described decay. As this decay spreads, more and more of the bone becomes exposed, in which respect also it differs from the former decay; and hence it may be called a denuding process. The bony substance of the Teeth also gives way, and the whole wasted surface has exactly the appearance, as if the Tooth had been filed with a rounded file, and afterwards had been finely polished. At these places the bony parts, being exposed, become brown.

I have seen instances, where it appeared as if the outer surface of the bony part, which is in contact with the inner surface of the enamel, had first been lost, so that the attraction of cohesion between the two had been destroyed; and as if the enamel had been separated for want of support, for it terminated all at once.

In one case, the two first incisors had lost the whole of the enamel; on their anterior surfaces, they were hollowed from side to side, as if a round file had been applied to them longitudinally, and had the finest polish imaginable. The three grinders on each side appeared as if a round file had been used on them, in a contrary direction to that on the incisors, *viz.* across their bodies close to the gum, so that there was a groove running across their bodies, which was smooth in the highest degree. Some of the other Teeth in the same jaw had begun to decay in a similar manner; also the Teeth in the lower jaw were become diseased.

I saw a case very lately, where the four incisors of the upper jaw had lost their enamel entirely on their anterior surfaces, and there was scarcely a Tooth in the mouth, which had not the appearance of having had a file applied across it close to the gum.

Those whom I have known, have not been able to attribute this disease to any cause; none of them had ever done anything particular to the Teeth, nor was there in appearance anything particular in the constitution, which could give rise to such a disease. In the first of these cases, the person was about forty; in the last, about twenty years of age.

From its attacking certain Teeth rather than others, in the same head, and a particular part of the Tooth, I suspect it to be an original disease of the Tooth itself; and not to depend on accident, way of life, constitution, or any particular management of the Teeth. (o)

(o) [The peculiar wearing away of the teeth, which Hunter has designated by the term Denudation, is most frequently confined to the six front teeth, but it may extend to the bicuspids and first molar. It usually commences at the necks of the teeth, and takes a horizontal direction, but Hunter is correct in saying that it may commence on the labial surfaces of the crowns, and that the enamel may be worn away in a longitudinal direction. The writer has seen the anterior surfaces of all the front teeth, upper and under, bevelled off, so that the crowns of the teeth were perfectly wedge-shaped with the narrow edge directed towards the interior of the mouth.

Mr. Bell has figured in his work on the teeth a case in which the greater portion of the crowns of the six anterior teeth was worn away, the destructive process having commenced on the cutting edge of the teeth and proceeded towards the necks until nearly half or one third of the crowns had disappeared. A similar case was under the notice of the writer for some years, and although the front teeth had not come in contact with each other since the commencement of the denuding process, the destructive action continued slowly to destroy the crowns until they were nearly level with the gums. It is also to be remarked that in both jaws the teeth were more acted upon towards the labial than the lingual surface, and the edges were rounded and not sharp.

Two theories have been put forth to account for this peculiar wearing away of the teeth; the one regards it as the result of a mechanical, the other of a chemical action. In the case recorded by Mr. Bell, and in the one last referred to, the side teeth of the upper and under jaws had not only come in direct contact with each other, but had worn their antagonists away, although not much beyond what is often met with in the mouths of many persons who are advanced in life, but as has been previously stated, a considerable space existed between the incisors and canines. Even supposing

§ 3. *Swelling of the Fang.*

Another disease of the Teeth is a swelling of the Fang, which most probably arises from inflammation, while the body continues sound, and is of that kind which in any other bone would be called a Spina Ventosa*. It gives considerable pain, and nothing can be seen externally.

The pain may either be in the Tooth itself, or the alveolar process, as it is obliged to give way to the increase of the Fang.

As a swelling of this kind does not tend to the suppurative inflammation, and as I have not been able to distinguish its symptoms from those of the nervous Tooth-ach, it becomes a matter of some difficulty to the operator; for the only cure yet known is the extraction of the Tooth; which has been often neglected on a supposition that the pain has been nervous.

---

* Vide Natural History of the Teeth, page 37.

---

that a mechanical force first excited the destructive action, it is difficult to understand how its influence should continue when the teeth no longer came in contact with each other. Again, if the action of the tooth-brush will serve to account for the wearing away in a horizontal direction at the necks of the teeth, it will not so readily explain the occurrence of those cases where the wearing away takes place in the length of the crown, still less will a mechanical theory explain the cases last mentioned.

In favour of the chemical theory may be mentioned the presence of numerous mucous glands in the substance of the lips, whose secretion may under certain circumstances have an acid reaction. The exposed line which is often produced at the neck of the tooth by the receding of the jaws will account for the usual situation of the denudation, but should the gum remain firmly adherent to the neck, then the secretion from the glands of the lips may begin to exact a destructive action on the most prominent parts of the teeth. As the edges of the front teeth become worn away, the lips naturally fall inwards, and continue to rest upon the anterior surface of the teeth, and might even fill up the space which we have seen is present between the six anterior teeth when the denudation has commenced at the cutting edges.

After all, either explanation is a matter of conjecture rather than of demonstration; much more attention is required to be bestowed upon the subject before it will be fully understood, but as the evidence stands at present, it appears to the writer to be in favour of the chemical theory.

These diseases of the Teeth, arising from inflammation, become often the cause of diseases in the alveolar processes, and gums; which I shall proceed to describe. (*p*)

(*p*) [The term *Spina ventosa* was first used by the Arabian writers to designate a disease in which matter formed in the interior of a bone, and then made its way to the surface. The following are Mr. Bell's remarks upon this section of Hunter's work:

"The affection here described is nothing more than a deposit of bony matter around the fang, produced, doubtless, by inflammation of the periosteum. The new bone is rather yellower and less opaque than the original structure. Hunter has alluded here to the occurrence of nervous pains, produced by these cases. It appears probable that the pressure of the new bone upon the nerves of the periosteum of the alveolus, or of the alveolar process itself, is the cause of this pain, for it cannot be distinguished from local neuralgia produced by any other similar cause. Cases of this description are detailed by Fox and by myself, in which the only means by which the true seat of pain could be ascertained was by striking the affected tooth, by which pain was produced, and the extraction of the tooth at once exhibited the cause of the pain and effected its cure." (1)

Such are the class of cases to which Hunter no doubt referred; but the writer has in his possession one specimen, presented to him by his friend Mr. W. Perkins, which far more closely corresponds to the meaning of the term *Spina ventosa*.

The tooth is a second lower molar belonging to the right side of the jaw; only a portion of the outer wall of the crown is left, the remainder having been destroyed by caries, leaving the continuation of the pulp cavity into the fangs exposed and open. The tooth was removed from the mouth of an elderly lady, having presented the usual symptoms which accompany chronic periostitis, with occasional paroxysms of a more acute character.

Upon examining the fangs they are seen to be somewhat enlarged, and the surface nodulated from ossific deposit. So far there is nothing to distinguish the tooth from any other that has been the subject of chronic periostitis, but on the anterior surface of the first fang near the apex is a small opening about the one-tenth of an inch in length, and the one-sixteenth in width, with the edges very jagged, and leading into a cavity excavated in the fang of the tooth. This cavity communicates by a narrow opening with the pulp cavity, as shown by fluid passing

---

(1) Hunter's Works by Palmer, vol. 2, p. 71.

§ 4. *Gum Boils.*

Although suppuration cannot easily take place within the cavity of a Tooth, yet it often happens, that the inflammation, which is extended beyond it, is so great, as to produce suppuration in the jaw at the bottom of the socket, where the diseased Tooth is, forming there a small abscess, commonly called a Gum Boil.

This inflammation is often very considerable, especially when the first suppuration takes place. It is often more diffused than inflammations in other parts, and affects the whole face, &c.

The matter, as in all other abscesses, makes its own way outwards, and as it cannot be evacuted through the Tooth, it destroys the alveolar processes, and tumifies the gum, generally on the fore part, either pointing directly at the root of the Tooth, or separating the gum from it; and is evacuated in one or other of these two ways, seldom on the inside of the gum; however, this sometimes happens.

Gum Boils seldom arise from other causes; however, it sometimes happens that they originate from a disease in the socket or jaw, having no connection with the Tooth, and only affecting it secondarily. Upon drawing such Teeth, they are generally found diseased at or near the point, being there very rough and irregular, similar to ulcerating bones. There is no disease to appearance in the body of the Tooth. These last described Gum Boils may arise wholly from such a cause, the appearance on the fang of the Tooth being only an effect.

These abscesses, whether arising from the Teeth or the sockets, always destroy the alveolar processes on that side where

---

from one to the other; but even if the formation of pus commenced in the pulp, which is not at all certain, the same morbid process must have extended itself to the interior of the dentine of the fang, and then found an exit at the opening on its anterior surface.]

they open; as is very evident in the jaw-bones of many sculls; on which account the Tooth becomes more or less loose. It may be perceived in the living body; for when the alveolar process is entirely destroyed on the outside of the Tooth, if that Tooth be moved, the motion will be observed under the gum, along the whole length of the fang.

So far the Teeth, alveolar processes, and gums, become diseased by consent.

It is common for these abcesses to skin over, and, in all appearance, heal. This is peculiar to those which open through the Gums, but those which discharge themselves between the Gums and Teeth, can never heal up, because the Gum cannot unite with the Tooth; however, the discharge in them becomes less at times, from a subsiding of the suppuration; which indeed is what allows the other to skin over. But either exposure to cold, or some other accidental cause, occasioning a fresh inflammation, produces an increased suppuration, which either opens the old orifice in the Gum, or augments the discharge by the side of the Tooth; however, I believe, the inflammation in this last case is not so violent as in the other, where a fresh ulceration is necessary for the passage of the matter.

Thus a Gum Boil goes on for years, healing and opening alternately; the effect of which is, that the alveolar processes are at length absorbed, and the Tooth gets looser and looser, till it either drops out, or is extracted.

Most probably in all such cases, the communications between the cavity of the Tooth, and the jaw, is cut off yet it keeps in part its lateral attachments, especially when the gum grasps the Tooth; but in those cases, where the matter passes between the Gum and the Tooth, their attachments are less; but some of them are still retained, particularly on the side opposite to the passage for the matter.

Gum Boils are easily known. Those which open through the Gum may be distinguished by a small rising between the arch of the Gum, and the attachment of the lip; upon pressing the Gum at the side of this point, some matter will commonly be observed, oozing out at the eminence. This eminence seldom subsides entirely; for even when there is no discharge, and the opening is healed over, a small rising may still be perceived, which shows that the Gum Boil has been there.

Those Gum Boils which discharge themselves between the Gum and the Tooth, are always discovered by pressing the Gum, whereby the matter is pressed out, and is seen lying in the angle between the Gum and Tooth. (*q*)

These abcesses happen much more frequently in the upper jaw, than in the lower, and also more frequently to the cuspidati, incisores, and bicuspides in that jaw, than to the molares; seldom to the fore Teeth in the lower jaw.

As Gum Boils are in general the consequence of rotten Teeth, we find them in young and middle aged people more frequently than in old; but they appear to be most common to the shedding Teeth. This will arise from those Teeth being more liable to become rotten; and perhaps there may be another reason, viz., the process of ulceration which goes on in these Teeth*, in some cases falling into suppuration. (*r*)

---

* Vide Natural History of the Teeth, for an Explanation of this process in those Teeth, pp. 140, 141.

(*q*) [In cases of rheumatic and some other forms of chronic periostitis a similar discharge may be seen to issue from around the neck of the tooth upon pressing the outer surface of the alveolus. It is very essential to distinguish these cases from ordinary gum-boil caused by caries. In the latter case it is almost invariably necessary to extract the tooth, while, in the former case, the disease is amenable to treatment.]

(*r*) [It does not appear that Hunter had any statistical details upon which he founded the above observations, inasmuch however as the six front teeth of the lower jaw are less liable to decay than the corresponding teeth of the upper, his statement is correct as regards the relative liability of these teeth to the formation of abscess, although of

It sometimes happens in these Gum Boils, that a fungus will push out at the orifice, from a luxuriant disposition to form granulations, in the inside of the abcess, and the want of power to heal or skin; the same thing frequently happens in issues, where the parts have a disposition to

---

course this does not absolutely prove the case, inasmuch as many teeth are lost from caries which yet never form an abscess or gum boil. Thus of 72 central incisors extracted between the ages of 15 and 60 years, 53 belonged to the upper jaw; of 117 lateral incisors 92 belonged to the upper jaw; and of 78 canines 58 belonged to the upper jaw. This same preponderance of numbers in the upper jaw exists as regards the bicuspids; thus of 273 first bicuspids removed between the same ages, 207 belonged to the upper jaw, and of 434 second bicuspids 279 belonged to the upper jaw. This relative liability of the loss of the teeth in the two jaws is however reversed when we come to the molars: thus, out of 1,124 first molars, 644 belonged to the lower jaw, out of 637 second molars, 388 belonged to the lower jaw, and out of 265 third molars 168 belonged to the lower jaw. These numbers are taken from the very valuable tables given by Mr. Tomes in his Lectures on Dental Physiology and Surgery.

The liability to the loss of certain teeth varies with the age of the individual as shown in the following Table taken from the same source.

TABLE.

Showing of 3,000 teeth extracted the per centage of the different kinds of teeth taken out between given ages, thereby showing that of teeth destined to be lost, the liability of the several kinds at each given age.

AGES.	TEETH.							
	Central Incisors.	Lateral Incisors.	Canines.	First Bicuspids.	Second Bicuspids.	First Molars.	Second Molars.	Third Molars.
	p. cent.	p. cent.	p. cent.	p. cent.	p. cent.	p. cent.	p. cent.	p. cent.
Under 15 - - -	2	3½	2½	7	8¾	68½	8	0
Between 15 and 20	1	2¾	1½	9½	16	44½	22¾	21½
,, 20 ,, 25	¾	2¾	¾	10¾	16½	34½	25	9½
,, 25 ,, 30	1½	2½	1	8	15	27	27	18
,, 30 ,, 40	2½	3½	3	10½	16½	22½	24	17¾
,, 40 ,, 50	7¼	8¾	7¾	9½	13	18½	19½	16
,, 50 ,, 60	8¼	9	3½	5½	14½	14	34½	10½
Upwards of 60	5½	13	8½	14	19¼	10¾	21½	7½

granulate, but have not the power of healing, on account of an extraneous body being kept there.

The Tooth in the present case acts as an extraneous body, and by the secretion of matter the abscess is prevented from healing.

In the treatment of Gum Boils, the practice will be the same, whether the abcess has arisen from a diseased Tooth, or a disease in the socket.

The Teeth being under such circumstances in the animal machine, that they cannot partake of all the benefits of a cure in the same manner as other parts do; on that account, when an abcess forms itself about the root of a Tooth, the Tooth by losing its connection with the other parts, loses every power of union, as it is not endowed with the power of granulating, and thereby it becomes an extraneous body, or at least acts here as an extraneous body, and one of the worst kind, such as it is not in the power of any operation of the machine to get rid of. This is not the case with any other part of the body, for when any other part becomes dead, the machine has the power of separating it from the living, called sloughing or exfoliation, and expelling it, whence a cure is effected; but in the case of Gum Boils, the only cure of them is the extraction of the Tooth. As this is the last resource, every thing is to be done to make the parts as easy under the disease as possible, so that this operation may be postponed.

When the abcess has opened through the Gum, I believe the best method that can be tried with a view to prevent future gatherings, is to prevent the closing up of the abcess; and this may be done, by enlarging the opening, and keeping it enlarged, till the whole internal surface of the cavity of the abcess is skinned over, or till the opening in the Gum loses the disposition to close up, which will in a great measure prevent any future formation of matter; or at least whatever is formed will find an easy outlet, which will prevent these accumulations from taking place ever after. The end of the fang will indeed be hereby exposed; but under such circumstances it will not be in a worse situation than when soaked in matter.

One method of doing this is to open the Gum Boils by a crucial incision, the full width of the abcess, and fill it well with lint, which should be dipped in lime water, or a diluted solution of lunar caustic; made by dissolving one drachm of the caustic in two ounces of distilled water; and the wound should be dressed very frequently, as it is with difficulty that the dressing can be kept in. If this is not sufficient to keep the wound open, it may be touched with the lunar caustic, so as to produce a slough; and this may be repeated, if it should be found necessary.

One considerable disadvantage occurs in this practice, which is the difficulty of keeping on the dressings; but constant attention will make up for the inconvenience of situation.

If the surface of the abcess be touched with the lapis septicus,[*] and the lip kept from coming in contact with the part for one minute, it answers better than any other method; for this, within that space of time, will penetrate to the bottom.

The surface of the boil should be first wiped dry, as much as the nature of the part will allow, to prevent as much as possible the spreading of the caustic; which by care can be prevented, as the operator will watch it the whole time.

To extract the Tooth, then to file off any diseased part of it, and immediately to replace it, has been practised, but often without the desired success; for it has often happened, that a Tooth has been introduced into a diseased jaw. This practice, however, now and then, has succeeded.

When a Gum Boil is formed on a back Tooth, or Molares,

---

[*] Potassa fusa.

such very nice treatment is not necessary, as when it happens to the fore Teeth; because, appearances are there of less consequence; therefore, the gum may be slit down upon the fang through its whole length, from the opening of the Gum Boil to its edge, which will prevent any future union; and the whole cavity of the abcess, skinning over, will prevent any future collection of matter. The wound appears afterwards like the hare lip, and therefore this practice is not advisable where it would be much in view; as when the disease is in the fore Teeth. In these cases, where the granulations push out through the small opening, they may be cured by the method above mentioned; but, if it is not complied with, they may be very safely cut off with a knife or lancet. However, this does not effect a cure; for they commonly rise again. To slit the gum, in his case has been common, but it is a bad method, whenever the defect is in sight. (*s*)

(*s*) [Alveolar abscess is the usual termination of acute inflammation of the dental periosteum. The progress of the inflammation and the prospect of effecting a cure depends first upon the nature of the exciting cause, and secondly upon the constitution of the patient. The most frequent cause is caries which having affected the pulp of the tooth the irritation extends to the membrane covering the fang, and sets up inflammation which terminates in the formation of pus. If seen at the commencement of the attack the object should be to arrest the inflammation before it has proceeded to the formation of pus, this may sometimes be accomplished by the application of a leech to the gum, and the employment of antiphlogistic treatment. When pus has formed, the most certain remedy is to extract the tooth, and the next best to make a free opening for the exit of the pus. Even when the abscess has been cured for the time it is very apt to form again and to be a perpetual source of trouble to the patient, nor can there be any prospect of a permanent cure in these cases until the caries has been first treated so as to prevent it from being a source of irritation for the future. It has been proposed of late years to freely open the alveolar abscess, and then to remove the secreting sac by the point of a bistoury. This it is said will sometimes effect a cure, but in other cases it is neces-

§ 5. *Excrescences from the Gum.*

From bad Teeth there are also sometimes excrescences, arising at once out of the Gum; near, or in contact with, the diseased Tooth.

---

sary afterwards to apply astringents or caustics to the part.—See 'Taft's Operative Dentistry,' p. 263.

Mechanical injury is another cause, when this happens the case must be treated upon the principles of general surgery, and if the immediate effects of the injury can be overcome then the tooth having been healthy previously there is every prospect of the inflammation subsiding, and the parts returning to their natural condition.

Inflammation of the dental periosteum sometimes accompanies a common cold, and when this happens, it generally yields to the same treatment as that which cures the cold.

Some time back, Dr. Maréchal de Calvi called attention to a peculiar form of dental periostitis, which he terms expulsive gingivitis. It bears a close resemblance to that condition described by Fox and others under the title of "scurvy in the gums." Dr. Edward Carrière says it is particularly prevalent where scrofula prevails. In the district of lower Austria, which extends from the south of the plains of Vienna, as far as the province of Styria, this disease attacks the teeth at all ages, and there are few of the inhabitants who have not suffered more or less from its ravages. The same authority states that its progress may be arrested by the use of the iodide of potassium. (1)

Dr. Graves, of Dublin, first clearly pointed out the existence of rheumatic periostitis in the membrane of the tooth, and showed that this also might be cured by the use of the iodide of potassium. See his Clinical Lectures.

When mercury has been given in excess it produces a peculiar kind of chronic periostitis, and unless the medicine is withdrawn, will speedily cause the loss of whatever teeth may remain in the mouth. This form of the disease is best combatted by the use of the chlorate of potash, taken internally and applied locally as a gargle. Dr. Watson recommends one part of brandy to four of water, to be used as a gargle in mercurial salivation. Tonic medicines and fresh air are also exceedingly beneficial in these cases.

In the three last forms of dental periostitis, the pus oozes from the necks of the teeth, and may be rendered visible by pressing on the gums over the affected teeth.]

---

(1) See L'Art Dentaire, vol. iv., also the writer's lectures on Diseases of the Dental Periosteum.

In general they are easily extracted with a knife, or whatever cutting instrument can be best applied; but this will vary according to their situations, and the extent of their base.

They will often rise in a day or two after the operation as high as ever; but this newly-generated matter generally dies soon, and the disease terminates well. They have often so much of a cancerous appearance, as to deter surgeons from meddling with them; but where they arise at once from the Gum, and appear to be the only diseased part, I believe they have no malignant disposition.

However, I have seen them with very broad bases, and where the whole could not be removed, and yet no bad consequences have attended their removal. These often rise again in a few years, by which means they become very troublesome.

After the extirpation of them, it is often necessary to apply the actual cautery to stop the bleeding; for arteries going to increased parts are themselves increased, and also become diseased; and have not the contractile power of a sound artery (*t*).

§ 6. *Deeply-Seated Abcesses in the Jaws.*

Sometimes deeper Abcesses occur than those commonly called Gum Boils. They are often of very serious consequences, producing carious bones, &c. These commonly arise from a disease in the Tooth, and more especially in the *cuspidati;* those Teeth passing farther into the jaw than the others. Their depth in the jaw being beyond the attachment of the lip to the gum, if an abcess forms at their

---

(*t*) [If the vascular excrescence depends upon the presence of a diseased tooth or the stump of a tooth, it will generally subside upon its removal; while every attempt at cure will fail, so long as the diseased tooth is allowed to remain.]

points, it more readily makes its way through the common integuments of the face, than between the gum and lip, which disfigures the face; and when in the lower jaw looks like the evil.

In the upper jaw it makes a disagreeable scar on the face about half an inch from the nose.

These, although they may sometimes arise from diseases of the Teeth and Gums, yet are properly the object of common surgery; and the Surgeon must apply to the Dentist, if his assistance is necessary, to pull out the Tooth, or to perform any other operation which comes under his province.

It sometimes happens that the abcess is situated some way from the root of the diseased Tooth, both in the upper jaw and the lower; but, I think, more frequently in the lower. (*u*) When it threatens to open externally on the skin of the face, great care should be taken to prevent it, and an opening very early made into the swelling on the inside of the lip; for it is generally very readily felt there. This

---

(*u*) [When an abscess is situated at a short distance from a tooth, the state of the gum around it, and the actual condition of the tooth will generally point to its true source. Harris, however, mentions a case where the patient had been troubled with dropping of pus from behind the curtain of the palate for about twelve months. Becoming alarmed, she consulted her physician, who satisfied himself that it arose from the socket of a diseased tooth; and after passing his finger around on the gums, covering the superior alveolar border, discovered a protuberance over the root of each upper central incisor, nearly as large as a hazel nut. Upon the removal of the teeth, the discharge of matter ceased. (1)

The same writer mentions another instance in which pus had escaped from the socket of a first superior molar, to about the centre of the palatine arch, thence passed up into the posterior nares, and was discharged from behind the velum palati.]

---

(1) The Principles and Practice of Dental Surgery. By Chapin A. Harris, M.D., D.D.S. 5th Edit., p. 465.

practice of early opening these abcesses upon the inside of the mouth is more necessary, when the abscess is in the lower jaw, than when in the upper; because matter by its weight always produces ulceration more readily at the lower part. I have seen this practice answer, even when the matter had come so near the skin, as to have inflamed it. If it is in the upper, the opening need not be so very large; as the matter will have a depending outlet.

To prevent a relapse of the disease, it will in most cases be necessary to pull out the Tooth; which has either been the first cause, or has become diseased, in consequence of the formation of the abcess; and in either case is capable of reproducing the disease.

The mouth should be often washed; and while the water is within the mouth, the skin should be pressed opposite to the abcess.

If the life of the bone be destroyed, it will exfoliate; and very probably two or three of the Teeth may come away with the exfoliation. Little should be done in such cases, except that the patient should keep the mouth as clean as possible by frequently washing it, and when the bones exfoliate, they should be removed as soon as possible. (*v*) In these cases it is but too common for the Dentist to be very busy, and perhaps do mischief through ignorance.

§ 7. *Abcess of the Antrum Maxillare.*

The *Antrum Maxillare* is very subject to inflammation and suppuration, by means of diseases of the neighbouring parts, and particularly of the duct leading to the nose being obliterated. Whether this is the cause, or only an effect, is

---

(*v*) [One of the best applications in all cases where there is diseased bone and an offensive discharge is, a gargle composed of the liquor sodæ chlorinatæ in the proportion one part to twelve of water. It acts as a deodoriser, and removes the offensive smell of the discharge.]

not easily determined, but there is great reason to suppose it an attendant, from some of the symptoms. If it be a cause, we may suppose, that the natural mucus of these cavities accumulating, irritates and produces inflammation for its own exit; in the same manner as an obstruction to the passage of the Tears through the *ductus ad Nasum*, produces an abcess of the *lachrymal sac*.

This inflammation of the *Antrum* gives a pain which will be at the first taken for the Tooth-ach, especially if there be a bad Tooth in that side; however, in these cases, the nose is more affected than commonly in a Tooth-ach.

The eye is also affected; and it is very common for people with such a disease to have a severe pain in the forehead, where the frontal sinus's are placed; but still the symptoms are not sufficient to distinguish the disease. Time must disclose the true cause of the pain; for it will commonly continue longer than that which arises from a diseased Tooth, and will become more and more severe; after which a redness will be observed on the fore part of the cheek, somewhat higher than the roots of the Teeth, and a hardness in the same place, which will be considerably circumscribed. This hardness may be felt rather highly situated on the inside of the lip.

As this disease has been often treated of by surgeons, I shall only make the following remarks concerning it.

The first part of the cure, as well as that of all other abcesses, is to make an opening, but not in the part where it threatens to point; for that would generally be through the skin of the cheek.

If the disease is known early, before it has caused the destruction of the fore part of the bone, there are two ways of opening the abcess; one by perforating the partition between the antrum and the nose, which may be done by drawing the first or second grinder of that side, and perforating the partition between the root of the alveolar process and the antrum, so that the matter may be discharged for the future that way.

But if the sore part of the bone has been destroyed, an opening may be made on the inside of the lip, where the abcess most probably will be felt; but this will be more apt than the other perforation to heal, and thereby may occasion a new accumulation; which is to be avoided, if possible, by putting in practice all the common methods of preventing openings from healing or closing up; but this practice will rather prove troublesome; therefore the drawing the Tooth is to be preferred, because it is not so liable to this objection. (*w*)

(*w*) ["Abscess of the antrum," says Druitt, in his 'Surgeons' Vade Mecum,' p. 443, 8th edit., "may be caused by blows on the cheek, but it more frequently results from the irritation of decayed teeth." Speaking of treatment, he observes, "A free aperture may be made into the cavity. If either of the molar teeth be loose or carious, it should be extracted, and a trochar be pushed through the empty socket into the antrum. But if all the teeth are sound, or if they have all been extracted before, an incision should be made through the membrane of the mouth, above the alveoli of the molar teeth, and the bone be pierced by a strong pair of scissors or trochar. The instruments should not be made of too highly tempered steel, lest they break. The cavity should be frequently syringed with warm water, in order to clear away the matter, which is sometimes thick, like putty. If the discharge continues profuse and fetid, search should be made with a probe for loose pieces of bone, which should be removed without delay, the aperture being enlarged if necessary."

Hunter appears to have confounded true abscess of the antrum, where pus is accumulated in the cavity, with dropsy of the antrum, which consists of an accumulation of the natural mucous secretion in consequence of the communication with the nostril having become closed up. The same distension of the antrum and thinning of its walls takes place as in the case of an abscess. Druitt quotes a case of this kind which occurred in the practice of Mr. Ferguson at King's College Hospital, in 1850, where there was great protrusion of the cheek and of the hard palate, and other signs of tumour, so that the patient was sent up for the purpose of having the bone extirpated; but on examination it was discovered that the antrum was greatly distended with a yellow viscid fluid, con-

CHAPTER II.

OF THE DISEASES OF THE ALVEOLAR PROCESSES, AND CONSEQUENCES OF THEM.

Having thus far treated of the diseases of the Teeth themselves, and those of the Sockets and Gums, which either arise from them, or are familiar to such as arise from them,

---

taining brilliant particles of cholesterine ; and an opening having been established through the anterior wall of the cavity, the patient was soon discharged cured.

The thinness of the walls of the antrum, which causes them to yield under pressure, and to give the crackling sound above-mentioned, would serve to diagnose abscess and dropsy of the antrum from the various kinds of solid tumours to which the same part is liable.

The condition of the teeth, the previous history of the case, and the absence of any offensive odour, will in most instances enable us to determine whether we have to deal with a case of abscess or dropsy of the antrum.

In the cases of abscess which have come under the notice of the writer, the accumulation of pus has originated from a diseased tooth, either a molar or bicuspid. The extraction of the tooth has been followed by a more or less free discharge of the matter from the antrum. It is, however, generally necessary to enlarge the opening from the socket of the extracted tooth, and to wash it out with tepid water. A piece of cotton-wool or lint, moistened with oil, should then be inserted, to prevent the closing up of the communication with the cavity while there is any discharge. Care should be taken that the plug is of such a size and shape as not to pass into the antrum. The discharge will usually subside after the removal of the tooth, unless the bone has become diseased.]

I come now to consider the diseases which take place primarily in the sockets, when the Teeth are perfectly sound: these appear to be two; and yet I am not sure but that they are both fundamentally the same, proceeding together from the same cause, or one depending on the other.

The first effect, which takes place, is a wasting of the Alveolar Processes, which are in many people gradually absorbed, and taken into the system. This wasting begins first at the edge of the socket, and gradually goes on to the root or bottom.

The Gum, which is supported by the Alveolar Process, loses its connection, and recedes from the body of the Tooth, in proportion as the socket is lost; in consequence of which, first the neck and then more or less of the fang itself, becomes exposed. The Tooth of course becomes extremely loose, and at last drops out.

The other effect is a filling up of the socket at the bottom, whereby the Tooth is gradually pushed out. As this disease seldom happens without being attended by the other, it is most probable that they generally both arise from the same cause. The second in these cases may be an effect of the first. Both combine to hasten the loss of the Tooth; but it sometimes happens that they act separately: for I have seen cases where the Gum was leaving the Teeth, and yet the Tooth was not in the least protruded; on the other hand, I have seen cases where the Tooth was protruding, and yet the Gum kept its breadth; but where this is not the case, and the Gums give way, the Gums generally become extremely diseased; and as they are separated from both the Teeth and the Alveolar Process, there is a very considerable discharge of matter from those detached surfaces.

Though the wasting of the Alveoli at their mouths, and the filling up at their bottoms, are to be considered as diseases when they happen early in life; yet it would appear to be only on account of a natural effect taking place too soon;

for the same thing is very common in old age ; * and also, as this process of filling up the bottom, and wasting of the mouths of the Alveolar Processes, takes place in all ages, where a Tooth has been drawn, and the connection between the two parts is destroyed this might lead us to suspect, that the original cause of these diseases may be a want of that perfect harmony, which is required between the Tooth and Socket, whereby a stimulus may be given in some degree similar to the loss of a Tooth; and by destroying that stimulus upon which the absorption of the process and the filling up of the socket depend, the natural disposition may be restored. This last opinion is strengthened by the following case.

One of the first incisors of the upper jaw of a young lady was gradually falling lower and lower. She was desirous of having a Tooth transplanted, which might better fit the shallow socket, as it was now become : she consulted me : I objected to this, fearing that the same disposition might still continue ; in which case the new Tooth would be probably pushed out in about half a year ; that the time since the old one began to sink, and a relief of so short a continuance, would be all the advantage gained by the operation ; but I observed at the same time, that the operation might have the effect of destroying the disposition to filling up, so that the new Tooth might keep its ground. This idea turned the balance in favour of the operation ; and it was performed. Time showed that the reasoning was just: the Tooth fastened and has kept its situation for some years.

These diseases arise often from visible causes. Any thing that occasions a considerable and long continued inflammation in those parts, such particularly as a salivation, will produce the same effect. The scurvy also, when carried to a great height, attacks the Gums, and the Alveolar Processes, which

---

* Vide 'Natural History of the Teeth,' page 8.

becomes a cause of the dissolution of those parts. This is most remarkable in the scurvy at sea.

When the disease arises from these two last causes, the Gums are either affected with the same disease together with the Alveolar Processes, or they sympathize with them. They swell, become soft and tender; and upon the least pressure or friction, bleed very freely.

How far these diseases can be prevented and cured, is, I believe, not known.

The practice hath been principally to scarify the Gums freely; and this with a view to fasten the Teeth made loose by the disease, which has therefore generally made a considerable progress before even an attempt towards a cure has been made. This scarification has certainly a good effect in some cases, the Teeth thereby becoming much faster; but how far the Alveolar Processes have been destroyed in such instances cannot be determined. Perhaps only a general fullness of the attaching membrane between the Tooth and the Process had taken place, as in a slight salivation, so as to push the Tooth a little way out of the bony socket; which having subsided by the plentiful bleeding, the Tooth of course becomes fast. Or perhaps, by producing an inflammation of another kind, the first inflammation or disposition to inflame is destroyed; which evidently appeared in the case of the young lady above mentioned.

If the above practice is unsuccessful, and the Tooth continues to protrude, it will either become very troublesome, or a great deformity. A fore Tooth may not, indeed, be at first so troublesome as a Grinder; because these Teeth frequently *overlop*; but it will be extremely disagreeable to the eye.

If the cause cannot be removed, the effect must be the object of our attention. To file down the projecting part is the only thing that can be done; but care must be taken not to file into the cavity, otherwise pain, inflammation, and other bad consequences, may probably ensue. This practice, however, will be very troublesome, because it will be difficult, to file a loose Tooth. At last the Tooth will drop out which will put an end to all farther trouble.

If the alveoli have really been destroyed, in those cases of loose Teeth which have become firm again, it would be difficult to ascertain whether they have a power of renewing themselves analogous to that power by which they first grow; or whether the fastening be effected by a closing of the Gum and Process to the Teeth. When the disease arises from the scurvy, the first attempt must be to cure that disease; and afterwards the above local treatment may be of service.

Together with drawing blood from the Gums, astringents have always been used to harden them. But when the disease does not arise from a constitutional cause, which may be removed, (such as the sea-scurvy or salivation) but from a disposition in the parts themselves, I have seen little relief given by them.

The tincture of myrrh, tincture of Peruvian bark, and sea-water, are some of the applications which have been recommended.

In such cases I have seen considerable benefit from the use of the tincture of bark and laudanum, in the proportion of two parts of the tincture of bark to one of the laudanum; and this to be used frequently, and at each time to be kept in the mouth during ten, fifteen or twenty minutes. (*x*)

(*x*) [The practice here recommended of filing a tooth which is descending in the alveolus is, on more than one account, exceedingly injurious. The immediate cause of such a descent of the tooth is, as Hunter observes, a deposit of bone in the alveolus; and, as this deposit takes place in consequence of some irritation in the periosteum, every thing should be avoided that could increase this irritation. Filing, however, would certainly increase it to a great degree. It would also tend greatly to lessen the attachment of the tooth to the socket, and the supports which the latter affords to it. The best mode of treatment appears to be, to apply leeches occasionally, particularly when there is

CHAPTER III.

OF THE DISEASES OF THE GUMS AND THE CONSEQUENCES OF THEM.

§ 1. *The Scurvy in the Gums.*

The Gums are extremely subject to diseases, the symptoms of which, in an advanced state of them, are in general such as were described in the preceding chapter.

They swell, become extremely tender, and bleed upon every occasion; which circumstances being somewhat similar to those observable in the true Scurvy, the disease has generally been called a Scurvy in the Gums.

But as this seems to be the principal way in which the Gums are affected, I suspect that the same symptoms may arise from various causes; as I have often seen the same appearances in children evidently of a scrofulous habit; and have also suspected them in grown people: they likewise frequently appear in persons, who are, in all other respects, perfectly healthy.

---

any unusual sensation produced by touching the tooth, indicating a degree of inflammation in the periosteum of the socket. This plan may be followed up by the use of astringent lotions. It is unnecessary to add, that any force applied to the tooth, and even frequently touching it, should be avoided. Ligatures are especially improper.— T. BELL.

These cases usually occur in persons about the middle period of life, and appear to depend upon some peculiar state of the constitution. There is but little hopes of arresting the progress of the disease by any local treatment.]

When the Gums first begin to have a tenderness, we may observe it first on their edges : the common smooth skin of the Gum is not continued to its very edge, but becomes at the edge a little rough like a border, and somewhat thickened. part of the Gum, between two Teeth, swells, and often pushes out like luxuriant flesh, which is frequently very tender.

The inflammation is often carried so far as to make the Gums ulcerate ; so that the Gums in many cases have a common ulcer upon them ; by which process, a part of the Teeth are denuded. This is often on one part only, often only on one jaw ; while in some cases it is on the whole Gums on both jaws.

In this case it often happens, that the Alveolar Process disappears, after the manner before described (see page 199) by taking part in the inflammation, either from the same cause or from sympathy. In such cases there is always a very considerable discharge of matter from the inside of the Gum, and Alveolar Process, which always takes the course of the Tooth for its exit.

In many of these cases we find that while the Gums are ulcerating in one part, they are swelling and becoming spongy in another, and hanging loose upon the Teeth ; and this often takes place, where there is no ulceration in any part.

The treatment, proper in this disease, where the Gums become luxuriant, from a kind of tumefaction, is generally to cut away all the redundant swellings of the Gum. I have seen several instances, where this has succeeded ; but still I am inclined to think, that this is not the best practice ; for it is not that an adventitious substance is thus removed, as in the case of luxuriant granulations, from a sore, but a part of the Gum itself is destroyed, in like manner as a part enlarged by inflammation may be reduced by the knife to its natural size, which would certainly be bad practice. I should suspect that the good arising from such practice, is owing to the bleeding which takes place; especially as I have found from experience, that simply scarifying the gums has answered the same purpose. Where there are reasons for supposing it to arise from a peculiarity in the constitution, the treatment should be such as will remove this peculiarity.

If the constitution is scorbutic, it must be treated with a view to the original disease. If scrofulous, local treatment, by wounding the parts, may do harm; but sea-bathing, and washing the mouth frequently with sea-water are the most powerful means of cure that I know. (*y*)

§ 2. *Callous Thickenings of the Gums.*

The Gums are also subject to other diseases, abstracted from their connection with the Alveoli and Teeth; which do not wholly belong to our present subject.

A very common one is the thickening of the Gum in some particular place, of a hard callous nature, similar to an excrescence. Many of these have a cancerous appearance, which deters the Surgeon from meddling with them; but in general without reason.

They may be often removed by the knife, but not always. The bleeding, which follows, is generally so considerable that it is frequently necessary to apply the actual cautery.

They sometimes grow again, which subjects the patient to the same operation. I have known them extracted six times; but, in such cases, I suspect that they really have a cancerous disposition; at least it has been so in two cases, which have fallen under my observation.

(*y*) [True scurvy is now seldom seen except in sailors. Cases of this kind which have occurred on board the Dreadnought Hospital Ship at Greenwich, have received the greatest benefit from the use of the chlorate of potash.]

But here the skill of the Surgeon, rather than that of the Dentist, is required. (*z*)

(*z*) [The most common enlargement to which the gums are subject is epulis. This consists of a fibrous or fibro-plastic growth of the gum. It generally commences between two teeth, but it may arise from the surface of the jaw-bone. These tumours are very apt to return after removal, and in order to effect a permanent cure, it is often necessary to remove the neighbouring teeth, when the sockets become absorbed, and the epulis, if confined to these parts, gradually disappears.]

CHAPTER IV.

OF NERVOUS PAINS IN THE JAWS.

There is one disease of the Jaws which seems in reality to have no connection with the Teeth, but of which the Teeth are generally suspected to be the cause. This deserves to be taken notice of in this place, because operators have frequently been deceived by it, and even sound Teeth have sometimes been extracted through an unfortunate mistake.

This pain is seated in some one part of the Jaws. As simple pain demonstrates nothing, a Tooth is often suspected, and is perhaps drawn out; but still the pain continues, with this difference however, that it now seems to be in the root of the next Tooth: it is then supposed either by the patient or the operator, that the wrong Tooth was extracted; wherefore, that in which the pain now seems to be, is drawn, but with as little benefit. I have known cases of this kind, where all the Teeth of the affected side of the Jaw, have been drawn out, and the pain has continued in the Jaw; in others, it has had a different effect, the sensation of pain has become more diffused, and has at last, attacked the corresponding side of the tongue.

In the first case, I have known it recommended to cut down upon the Jaw, and even to perforate and cauterise it, but all without effect.

Hence it should appear, that the pain, in question, does

not arise from any disease in the part, but is entirely a nervous affection.

It is sometimes brought on, or increased, by affections of the mind, of which I once saw a remarkable instance in a young Lady.

It often has its periods, and these are frequently very regular.

The regularity of its periods gives an idea of its being a proper case for the bark, which, however, frequently fails.

I have seen cases of some years standing, where the hemlock has succeeded, when the bark has had no effect; but sometimes all attempts prove unsuccessful. Sea bathing has been in some cases of singular service. (*a*)

(*a*) [Neuralgia of the fifth pair of nerves may arise from teeth that are perfectly sound, so far as the presence of caries is concerned, but where there is an increased growth of the cementum around the fang, producing exostosis. In these cases, the patient will point to a particular tooth as the seat of the pain. If the disease has lasted for some time, and no relief has been obtained by other means, and supposing it is quite clear that the neuralgia does not depend upon any of the causes presently to be mentioned, it would be justifiable to remove the tooth. If an exostosis is found on the fang, the operation will most probably be followed by temporary relief, but, unfortunately, in these cases several of the remaining teeth are often affected in the same way, and the pain recurs after a short interval until all the teeth are removed. Mr. Fox mentions a case in which a lady lost all her teeth before she obtained a cessation of her sufferings, which arose from exostosis, and Mr. Tomes has recorded a similar case. (1)

Hysterical females will sometimes complain of neuralgia of the head and face, pointing to a perfectly healthy tooth as the source of their sufferings, and requesting to have it removed. The condition of the tooth, the age, and appearance of the patient, and the general history of the case, will enable us to diagnose the true character of the affection.

Malaria is another cause of this complaint, and here the character of

---

(1) 'System of Dental Surgery,' pp. 441.

CHAPTER V.

OF THE EXTRANEOUS MATTER UPON THE TEETH.

There are parts of the Tooth, which lie out of the way of friction, viz. the angles made by two Teeth, and the small indentation between the Tooth and Gum.

Into these places the juices are pressed and there stagnate, giving them at first the appearance of being stained or dirty A Tooth in this stage is generally clean for some way from its cutting edge, towards the gum, on account of the motion of the lips upon it, and the pressure of the food, &c. It is also pretty clean close to the Gum, from the motion of the loose edge of the Gum upon that part, but this circumstance is only observable in those who have their Gums perfectly sound; for in others, this loose edge of the Gum is either lost, or no longer retains its free motion.

If art be not now used, as the natural motion of the parts is not sufficient, the incrustation increasing covers more and more of the Teeth. As mastication generally

---

the district in which the patient resides, and the history of the complaint will afford a clue to its real nature. It is in these cases that quinine will often give relief. Disease at the base of the brain, or in the course of the trunk of the nerve, may constitute another cause of facial neuralgia. These cases are sometimes very difficult to diagnose, while at other times they are evidently accompanied by disease of the brain.]

keeps that part clear which is near to the edges and grinding surfaces; and as the motion of the lips in some measure retards its growth outwards, it accumulates on the parts above mentioned, till it rises almost as high as the Gum; its growth being now retarded in that direction, it accumulates on the edge next to the Gum, so that in time it passes over the Gum, of which it covers a greater or less portion. When it has encreased so much as to touch the Gum, (which very soon happens especially in the angle between the Teeth) it produces ulceration of that part, and a train of bad consequences. Often the Gums, receding from this matter, become very tender and subject to hemorrhage.

The Alveolar Processes frequently take part with the Gums, and ulcerate, so that the Teeth are left without their support, and at last drop out, similarly to the diseases of these parts already described.

All our juices contain a considerable quantity of calcareous earth, which is dissolved in them, and which is separated from them upon exposure, which continues mixed with the mucus; so that the extraneous matter consists of earth and common secreted mucous.*

This disposition of the juices of the mouth to abound so much with earth, seems to be peculiar to some people, perhaps to some constitutions; but I have not been able to ascertain what these are. We find persons who seem to have nothing particular, either in constitution or way of life, so subject to this accumulation, that the common methods of prevention, such as washing and brushing the Teeth, have not the desired effect.

The disposition is so strong in some people, that the concretion forms on the whole body of the Tooth; I have

---

* Vide Natural History, page 155, in the Note, for a further description of this.

seen it even on the grinding surface of the *molares*, and often two or three Teeth are cemented together with it. This I think could only happen to those who seldom or never use these Teeth. It is very apt to accumulate on a Tooth the opposite of which is lost.

I once saw a case of this kind, where the accumulation, which was on a grinder, appeared like a tumour on the inside of the mouth, and made a rising in the cheek, which was supposed by every one that felt it, to be a scirrhous tumour forming on the cheek; but it broke off and discovered what it was.

This accumulation is very apt to begin during a fit of sickness, when the extraneous juices are allowed to rest; and perhaps the juices themselves may have at this time a greater tendency to produce the encrusting matter.

It may also arise from any circumstance, which prevents a person from eating solids, whereby the different parts of the mouth have less motion on each other. Lying-in women are instances of this; not to mention that the assistance of art in keeping the Teeth clean is commonly wanting under such circumstances.

The adventitious substance, as was said before, is composed of mucous or animal juices, and calcareous earth; the earth is attached to and crystalized upon the Tooth, and the mucus is intangled in these crystals.

The removal of this adventitious matter, is a part in which the dentist ought to be very cautious; he should be perfectly master of the difference between the natural or original Tooth, and the adventitious matter; and he should be sensible of the propriety of saving as much as possible of the Tooth, and at the same time take pains to remove all that which is not natural. Many persons have had their teeth wholly spoiled by an injudicious treatment of them in this respect.

As the cause of this incrustation is not either a known disease of the constitution, or of the parts but depends on a property of the matter, secreted, simply as inanimate matter; the remedy of course becomes either mechanical or chemical.

The mechanical remedies are friction, filling and picking. The first is sufficient, when the Teeth are only beginning to be discoloured; or, when already clean, they may be thus kept clean. Various are the methods proposed; to wash them with cold water, and at the same time to rub them with a piece of cloth on the fore-finger, has been thought sufficient by some; others have recommended the dust of a burnt cork, burnt bread, &c. with a view to act with more power on the adventitious matter, than what can be applied by the means of a soft brush or cloth.

In cases where this incrustation has been more considerable, powders of various kinds have been employed, such as tartar, bole, and many others.

Cream of tartar is often used, which at the same time that it acts mechanically, has likewise a chymical power, and dissolves this matter.

Other mechanical means are instruments to pick, scrape and file off the calcareous earth; these should only be made use of when it is in large quantities, and with great caution, as the Teeth may be somewhat loose; or, a part of the Tooth may be broken off with the incrustation.

The chymical means are solvents: these are either alkalies or acids; the alkaline salt will answer very well early in the disease; for the crust of the first stage consists chiefly of mucus, which the alkali will remove very readily: but it should not be used too freely, as it rather softens the Gums, and makes them extremely tender.

Acids are also employed with success, as they dissolve the earth, but are attended with this disadvantage, that they act with more force upon the Tooth itself, dissolving part of it, which is to be avoided, if possible; for no part of a sound Tooth can be spared.

We may observe that people who eat a good deal of salad or fruit, have their Teeth much cleaner than common; which is owing to the acids in those fruits; and for the same reason people's Teeth are commonly cleaner in summer than winter, in those countries where there is a great plenty of fruit. When the accumulation has been considerable, the Teeth and Gums will feel tender on the removal of this matter, and even be affected by cold air; but this will not be of long continuance. (*b*)

(*b*) [According to Berzelius, tartar consists of—

Earthy phosphates	79·0
Salivary mucus	12·5
Ptyalin	1·0
Animal matter soluble in hydrochloric acid . . .	7·5

This deposit is derived from the mixed saliva of the mouth; it collects principally behind the lower incisors and canines, and between the first and second molar teeth of the upper jaw. The first situation is the most dependent part of the mouth; it is were the saliva accumulates, and it is at this part that the ducts of the salivary and submaxillary glands open. The duct of the parotid gland opens opposite the second molar of the upper jaw. In persons who have a great tendency to the formation of tartar, it collects on all the teeth, but still to a much greater extent in localities mentioned than elsewhere.

Removal of the tartar by proper instruments, and by the use of a moderately coarse powder, which exerts no chemical action either on the deposit or on the teeth, are the only means which should be used. All acids must be avoided, since what would act upon the tartar would also affect the teeth. When the gums are spongy, vascular, or ulcerated, after removing the tartar, they should be freely lanced for the purpose of unloading the vessels, and an astringent gargle used two or three times in the course of the day.]

CHAPTER VI.

OF THE IRREGULARITY OF THE TEETH.

As that part of each jaw, which holds the ten fore-teeth, is exactly the same size when it contains those of the first set, as when it contains those of the second; and as these last often occupy a much larger space than the first*, in such cases the second set are obliged to stand very irregularly.

This happens much oftner in the upper-jaw, than in the lower; because, the difference of size of the two sets is much greater in that jaw.

This irregularity is observed almost solely in the incisors and cuspidati; for they are the only Teeth which are larger than their predecessors.

It most frequently happens to the cuspidati, because they are often formed later than the bicuspides; in consequence of which the whole space is taken up before they make their appearance: in such cases they are obliged to shoot forwards or outwards over the second incisor. However, it frequently happens to the incisors, but seldom to such a degree. This arises often from the temporary cuspidatus of one or both sides standing firm. I have seen the irregularities so much as to appear like a double row.

The bicuspidati generally have sufficient room to grow,

---

* Vide Natural History, pages 141, 142.

because even more space, than what they can occupy is kept for them by the temporary grinders.* This however, is not universally the case; for I have seen where the bicuspidati were obliged to grow out of the circle, very probably from their being later in growing than common.

That it is from want of room in the jaw, and not from any effect that the first set produce upon them is evident; first, because in all these cases of irregularity we find that there is really not room in the Jaw, to allow of placing all the Teeth properly in the circle; so that some are necessarily on the outside of the circle, others within it, while others are turned with their edges obliquely as it were, warped; and secondly, because the bicuspides are not out of the circle, although they are as much influenced by the first set as any of the others.

As they are not influenced by the first set, it cannot be of any service to draw the first possessor; for that gives way in the same proportion as the other advances. As the succeeding Tooth however is broader, it often interferes with a shedding Tooth next to it, the fang of which not being influenced by the growth of its own succeeding Tooth, it does not decay in proportion as the other advances, and therefore the drawing of the adjoining shedding Tooth is often of service.†

In cases of considerable irregularity for want of room, a principal object is to remove those which are most out of their place, and thereby procure room for the others which are to be brought into the circle.

To extract an irregular Tooth would answer but little purpose, if no alteration could be made in the situation of the rest; but we find that the very principle upon which Teeth are made to grow irregularly, is capable, if properly

---

* Vide Natural History, pages 121, 127.
† Ibid, 143, 144.

directed, of bringing them even again. This principle, is the power which many parts (especially bones) have of moving out of the way of mechanical pressure.

The irregularity of the Teeth is at first owing to mechanical pressure; for one Tooth getting the start of another, and fixing firmly in its place, becomes a resistance to the young, loose, forming Tooth, and gives it an oblique direction. The same principle takes place in a completely formed Tooth, whenever a pressure is made upon it. Probably a Tooth might by slow degrees be moved to any part of the mouth, for I have seen the cuspidati pressed into the place of the incisores. However it is observed, that the Teeth are easier moved backwards than forwards, and when moved back that they are permanent, but often, when moved forwards, that they are very apt to recede.

The best time for moving the Teeth is in youth, while the jaws have an adapting disposition; for, after a certain time, they do not so readily suit themselves to the irregularity of the Teeth. This we see plainly to be the case, when we compare the loss of a Tooth at the age of fifteen years, and at that of thirty or forty. In the first case we find, that the two neighbouring Teeth approach one another, in every part alike, till they are close: but in the second, the distance in the jaw, between the two neighbouring Teeth, remains the same, while the bodies will in a small degree incline to one another for want of lateral support.

And this circumstance of the bodies of the Teeth yielding to pressure upon their base, shows that, even in the adult, they might be brought nearer to one another by art properly applied.

As the operation of moving the Teeth is by lateral pressure upon their bodies, these bodies must first have passed through the gum sufficiently for a hold to be taken.

The best time seems to be, when the two grinders of the child have been shed; for at this time a natural alteration is taking place in that part of the jaw.

The means of making this pressure I shall only slightly

describe, as they will greatly vary according to circumstances, so considerably indeed, that scarcely two cases are to be treated alike, and in general the dentists are tolerably well acquainted with the methods.

In general, it is done with ligatures or plates of silver. The ligatures answer best when it is only required to bring two Teeth closer together, which are pretty much in the circle. The trouble attending this is but trifling, as it is only that of having them tied once a week or fortnight.

Where Teeth, growing out of the circle, are to be brought into it, curved silver plates, of a proper construction must be used. These are generally made to act upon three points, two fixed points on the standing Teeth, and the third on the Tooth which is to be moved. That part of the plate, which rests on the two standing Teeth, must be of a sufficient length for that purpose, while the curved part is short, and goes on the opposite side of the Tooth to be moved. Its effect depends very much on the attention of the patient, who must frequently press hard upon it with the Teeth of the opposite jaw; so that this method is much more troublesome to the patient than the ligature.

It is impossible to give absolute directions what Tooth or Teeth ought to be pulled out. That must be left to the judgment of the operator; but the following general hints may be of service.

1. If there is any one Tooth very much out of the row, and all the others regular, that Tooth may be removed, and the two neighbouring ones brought closer together.

2. If there are two or more Teeth of the same side very irregular, (as for instance, the second incisor and cuspidatus) and it appears to be of no consequence, with respect to regularity, which of them is removed, I should recommend the extraction of the farthest back of the two, viz., the cuspidatus; because, if there should be any space, not filled up, when the other is brought into the row, it will not be so readily seen.

3. If the above-mentioned two Teeth are not in the circle, but still not far out of it, and yet there is not room for both; in such a case I would recommend the extraction of the first bicuspis, although it should be perfectly in the row, because the two others will then be easily brought into the circle; and, if there is any space left, it will be so far back as not to be at all observable.

The upper jaw is often rather too narrow from side to side, near the anterior part which supports the fore Teeth, and projects forwards considerably over the lower, giving the appearance of the rabbit-mouth, although the Teeth be quite regular in the circle of the jaw.

In such a case it is necessary to draw a biscuspis of each side, by which means the forepart of the circle will fall back; and if a cross bar was to be stretched from side to side across the roof of the mouth, between cuspis and cuspis, it would widen the circle. The fore Teeth might also be tied to this bar, which would be a means of assisting nature in bringing them back. This has been practised, but it is troublesome.

As neither the bodies nor the fangs of the Teeth are perfectly round, we find that this circumstance often becomes a cause of their taking a twist; for, while growing, they may press with one edge only on the completely formed Tooth, by which means they will be turned a little upon their center.

The alteration of these is more difficult than of the former, for it is, in general, impossible to apply, so long and constantly as is necessary for such an operation, any pressure that has the power of turning the Tooth upon its center. However, in the incisores, it may be done by the same powers which produce the lateral motion; but where these cannot be applied, as is frequently the case, the Tooth may be either pulled out entirely, and put in again even, or it may be twisted round sufficiently to bring it into a proper position, as hath been often practised.

It may not be improper, in this place, to take notice of

a case which frequently occurs. It is a decay of the first adult grinder at an early age, viz. before the temporary grinders are shed, and before the second grinder of the adult has made its appearance through the gum. In this case, I would recommend removing the diseased Tooth immediately, although it may occasion no kind of trouble; for if it be drawn before the temporary grinders are shed, and before the second adult grinder has cut the gum, it will in a short time not be missed; because the bicuspis of that side will fall a little back, and the second and third grinders will come a little forward; by which means the space will be filled up, and these Teeth will be well supported. Besides, the removal of this Tooth, will make room for the fore Teeth, which is very often much wanted, especially in the upper jaw.

CHAPTER VII.

OF IRREGULARITIES BETWEEN THE TEETH AND JAW.

Certain disproportions between the Teeth and Jaw, sometimes occur, one of which is, when the body of the lower jaw is not of sufficient length for all the Teeth. In such cases, the last grinder never gets perfectly from under the coronoide process, its anterior edge only being uncovered; and the gum, which still in part lies upon the Tooth, is rubbed against the sharp points of the Tooth, and is often squeezed between the Tooth upon which it lies, and the corresponding one of the upper jaw. This occasions so much uneasiness to the patient, that it becomes necessary to relieve the gum, if possible by dividing it freely in several places, that it may shrink and leave this surface of the Tooth wholly uncovered. If this does not answer, which is sometimes the case, it is adviseable to draw the Tooth.

Sometimes, although but seldom, an inconvenience arises from the dentes sapientiæ being in the upper Jaw, and not in the lower; these Teeth pressing upon the anterior part of the root of the coronoide process when the mouth is shut, for the coronoide processes are farther forwards in such cases, than when the lower Jaw also has its dentes sapientiæ; in short, the exact correspondence between the two Jaws is not kept up.

In such cases I know of no other remedy but the extraction of the Tooth.

OF SUPERNUMERARY TEETH.

When there are Supernumerary Teeth,* it will, in general, be proper to have them drawn; for they are commonly either troublesome, or disfigure the mouth.

---

* Vide Natural History, page 143.

CHAPTER VIII.

OF THE UNDER JAW.

It is not uncommon to find the lower Jaw projecting too far forwards, so that its fore Teeth pass before those of the upper Jaw, when the mouth is shut;* which is attended with inconvenience, and disfigures the face.

This deformity can be greatly mended in young people. The teeth in the lower Jaw can be gradually pushed back in those, whose Teeth are not close, while those in the upper can be gently brought forward; which is by much the easiest operation.

These two effects are produced by the same mechanical powers. While this position of the Jaw is only in a small degree, so that the edges of the under Teeth can be by the patient brought behind those of the upper, it is in his own power to encrease this, till the whole be completed; that is till the grinders meet; and it is not necessary to go farther. This is done by frequently bringing the lower Jaw as far back as he can, and then squeezing the Teeth as close together as possible.

But when it is not in the person's power to bring the lower Jaw so far back, as to allow the edges of its fore Teeth to come behind those of the upper, artificial means are necessary.

---

* Vide Natural History, p. 108.

The best of these means is an instrument of silver, with a socket or groove shaped to the fore Teeth of the lower jaw to receive them, so as to become fast to them, and sloped off as it arises to its upper edge, so as to rise behind the fore Teeth in the upper jaw in such a manner, that upon shutting the mouth, the Teeth of the upper jaw may catch the anterior part of the slanting surface, and be pushed forward with the power of the inclined plane. The patient, who wears such an instrument, must frequently shut his mouth with this view.

These need not be continued longer than till the edges of the lower Teeth can be got behind those of the upper; for it is then within the power of the patient, as in the first stated case. (c)

(c) [Hunter's remarks upon the treatment of irregularities of the teeth are sound and judicious as far as they go, although his ideas with respect to the extraction of the temporary teeth are evidently influenced by his opinion that the ten permanent teeth only require exactly the same amount of space as that which was occupied by their predecessors.

The treatment of irregularities has made such advances since Hunter's time, that it is impossible to describe the various contrivances now employed, without exceeding the limits of notes. The reader is therefore referred to the standard works on Dental Surgery for information on this part of the subject.]

CHAPTER IX.

OF DRAWING THE TEETH.

The extraction of Teeth, is, in some cases, an operation of considerable delicacy, and, in others, no operation is less difficult.

As this is often not thought of till an inflammation has come on, it becomes an object of consideration whether it be proper to remove the Tooth while that inflammation continues, or to wait till it has subsided. I am apt to believe it is better to wait even till the parts have perfectly recovered themselves, because the state of irritation renders them more susceptible of pain. The contrary practice might also appear reasonable, for by removing the Tooth it might be imagined that we should remove the cause; but when the inflammation has once begun, the effect will go on independently of the cause; and to draw the Tooth, in such a situation, is rather to produce a fresh cause, than to remove the present. Of this I think an instance has occurred to me. However, most teeth are drawn in the height of inflammation: and, as we do not find any mischief from the operation, it is perhaps better to do it when the resolution of the patient is the greatest. The sensibility of the mind may even be less at this time.

Teeth are easy or difficult of extraction, according as they are fast or loose in their sockets; in some degree according to the kind of Tooth, and also, in some degree, with reference to their situation.*

---

* For farther directions, vide Natural History, page 154.

They are naturally so fast as to require instruments; and the most cautious and dexterous hand; and yet are sometimes loose enough to be pulled out by the fingers.

When the sockets and gums are considerably decayed, and the Tooth or Teeth very loose, it would in most cases be right to perform extraction; for when they are allowed to stay, and perhaps are kept in their proper place by being tied to the neighbouring Teeth, they then act upon the remaining gum and socket as extraneous bodies, producing ulceration there, and making those parts recede much farther than they naturally would have done, if the Tooth had been drawn earlier; which produces two bad effects, it weakens the lateral support of the two neighbouring Teeth, and it renders it more difficult to fix an artificial Tooth. But unless these two last circumstances are forcibly impressed upon the patient, it is hardly possible to persuade him to consent to the loss of a Tooth while it has any hold, especially a tooth which appears sound.

The extraction should never be done quick; for this often occasions great mischief, breaking the Tooth or jaw; on the same principle, as a bullet, going against an open door with great velocity, will pass through it, but, with little velocity, will shut it.

This caution is most necessary in adults, or in the permanent Teeth*; for in young subjects, where there are only the temporary Teeth†, the jaw, not being so firm, the Tooth is not in much danger of being broken.‡

It is a common practice to divide the gum from the Tooth

---

* Vide Natural History, page 124.
† Vide Natural History, page 140.
‡ I must do Mr. Spence the justice to say, that this method appears to be peculiar to him, and that he is the only operator I ever knew, who would submit to be instructed, or even allow an equal in knowledge; and I must do the same justice to both his sons.

before it is drawn, which is attended with very little advantage; because at best it can only be imperfectly done, and that part of the gum, which adheres to the Tooth, decays when it is lost. But if such a separation, as can be made, saves any pain in the whole of the operation, I should certainly recommend it; and at least in some cases, it might prevent the gum from being torn. It is also a common practice to close the gum as it is termed; this is more for show than use; for the gum cannot be made so close as to unite by the first intention; and therefore the cavity from which the Tooth came, must suppurate like every other wound. But, as the sensations of these parts are adapted to such a loss; and, as a process very different from that which follows the loss of so much substance in any other part of the body, is to take place; the consequent inflammations and suppurations are not so violent.* We may be allowed to call this a natural operation which goes on in the gum and alveoli, and not a violence; as we see that the delivery of a young animal before its time, which is similar to the drawing of a fixed Tooth, in happening before all the containing parts are prepared for the loss, produces considerable local violence, without doing proportionable mischief. Therefore, in general, it is very unnecessary to do any thing at all to the gum.

There are some particular circumstances, which naturally, and others which accidentally attend and follow the drawing of Teeth; but they are in general of no great consequence.

There follows a bleeding from the vessels of the socket, and those passing between it and the Teeth.† This commonly is but trifling; however, instances have occurred where it has been very considerable, and the awkwardness

---

* Vide Natural History, on decay of the Alveoli, page 8.
† Vide Natural History, pp. 53, 54, 55.

of the situation makes it very difficult to stop it. In general it will be sufficient to stuff the socket with lint, or lint dipped in the oil of turpentine, and to apply a compress of lint or a piece of cork thicker than the bodies of the adjacent Teeth, so that the Teeth in the opposite jaw may keep up a pressure.

It has been advised to stuff into the socket some soft wax, on a supposition that it would mould itself to the cavity, and so stop the bleeding; this perhaps may sometimes answer better than the other method, and therefore should be tried when that fails. (*d*)

(*d*) [Hæmorrhage, after the extraction of a tooth, is seldom so severe as to excite alarm, excepting when the hæmorrhagic diathesis is present. In these cases death has sometimes ensued. The following is a summary of the principal cases I have found recorded of fatal or of dangerous hæmorrhage from the extraction of a tooth. Plater mentions the case of a man who died from this cause in 1555; Schenck has recorded another instance, and M. Courtois gives the history of a third case in the *Dentiste Observateur*, 1775. The patient was scrofulous, and the bleeding came rather from the gums than the tooth. Coming down to cases of more recent date, I find Snell mentions the death of an elderly gentleman who had one loose tooth, a bicuspis, which was extracted; hæmorrhage ensued, and in spite of remedies he died on the day after the operation. In the eighth volume of the 'Med. Chir. Transactions,' is the history of a fatal case which occurred in the practice of Mr. Blagden. The patient was a male, twenty-six years of age; the tooth from which the bleeding came, was a second molar of the upper jaw. The actual cautery was applied, and the carotid artery tied. The hæmorrhagic diathesis was present, and the patient died on the sixth day.

In the 'Medical Gazette' of 1841, Mr. Robertson, of Edinburgh, has given the history of a fatal case in a gentleman of middle age, after the extraction of a loose wisdom tooth in the lower jaw. He was of a full habit of body, the hæmorrhagic diathesis was present, and he died on the twenty-third day. The actual cautery was applied with only temporary benefit, and the lip having been burnt in the operation, this afterwards became a fresh source of bleeding; so also in Mr. Blagden's case the wound produced in tying the carotid artery, afterwards bled and hastened the fatal result.

In a case related by Mr. Davenport in the 'Medical Gazette' for 1862,

It is scarcely possible to draw some Teeth without breaking the Alveolar Processes. This in general is but of little

---

the patient was a healthy man, twenty-three years of age, and the tooth a molar in the lower jaw. The bleeding continued for thirty-four hours, and then stopped; a large quantity of blood was lost, and the patient was in considerable danger.

Dr. Richardson met with an instance in which hæmorrhage occurred in a man after extraction of one of the lower molars, and continued for four days. The man was exhausted, and the blood was coming away freely from the socket of the tooth. It was arrested by the application of dilute nitric acid and pledgets of lint, after thoroughly clearing out the socket, by syringing it with warm water. (1)

Mr. Mason Good, in his 'Study of Medicine,' has given the following instructive case, showing the danger into which a patient may be brought from the improper application of the means usually employed for arresting hæmorrhage after tooth extraction :—" I was, not long ago, requested to see a young man who had been profusely bleeding from the gum and socket of an extracted tooth five days without cessation and without sleep, till his wan cheeks and faint emaciated frame seemed to indicate that he had scarcely any blood left in his vessels. He was so weak as to be incapable of rising from his bed or taking food, and his stools, from the quantity of blood he was perpetually swallowing, had all the appearance of a melœna. On opening his mouth, I found it crammed full of lint and wadding, one piece having every hour been added to another without a removal of the preceding, lest the hæmorrhage should be increased; whilst the blood in which the wadding was soaked, and which had remained in the socket and over the gums for so long a period, was become grumous, putrid, and intolerably offensive.

"I first removed the whole of this nauseating load from the patient's mouth, and gave him some warm brandy-and-water to wash it with. I next directed him to take a goblet of negus, with a little biscuit sopped in it, a part of which he soon contrived to swallow. The bleeding still continued; but, as I had little doubt that this proceeded entirely from a want of power in the lacerated arteries to contract, I applied no pressure of any kind, but prescribed a gargle of equal parts of tincture of catechu and warm water, and the hæmorrhage soon ceased."

From the consideration of these cases, we learn the following important facts: First, wherever death has ensued, and we possess a full knowledge of the circumstances, the patient has had the hæmorrhagic

---

(1) *On the Medical History and Diseases of the Teeth.* By W. B. Richardson, M.A., M.D. P. 45. Lond., 1860.

consequence, because from the nature of the union between the Teeth and sockets, these last can scarcely be broken farther than the points of the fang, and in very few cases so far; therefore little mischief can ensue, as the fracture extends no farther than the part of the socket which will naturally decay after the loss of the Tooth; and that part which does not decay, will be filled up as a basis for the gum to rest upon. It has been supposed that the splinters do mischief. I very much doubt this; for if they are not so much detached as to lose the living principle, they still continue part of our body, and are rounded off at their points as all splinters are in other fractures, and particularly here, for the reasons already assigned, viz., because this part has a greater disposition for wasting. And if they are wholly detached, they will either come away before the gum contracts entirely; or, after it is closed, will act as an extraneous body; form a small abcess in the gum; and come out.

---

diathesis, or was the subject of scurvy: in the latter case, the bleeding did not come so much from the socket of the tooth as from the surface of the gums. Secondly, the bleeding lasts for several days, and ample time is allowed for the administration and operation of internal remedies.

In cases where the hæmorrhagic diathesis is present, we ought, therefore, to rely rather upon constitutional (1) than local remedies, but when the patient is plethoric, or only of a weak constitution, local remedies may generally be depended upon to arrest the bleeding.

Of the various styptics that have been recommended to arrest the bleeding, the most valuable are the tinctura ferri, sesquichloridi and creosote.

With regard to creosote, it is necessary to remark that the kind commonly met with acts merely as a mild astringent; but that known as German creosote acts as a powerful caustic, and then it is necessary to be careful that none of it flows on to the other parts of the mouth. Mialhe regards it as one of the most powerful coagulants of albumen that we possess.]

---

(1) See Richardson, *opus. cit.*

It sometimes happens, that the Tooth is broken, and its point, or more of the fang is left behind, which is very often sufficient to continue the former complaints; and therefore it should be extracted, if it can be done, with care. If it cannot be extracted, the gum will in part grow over it; and the Alveoli will decay as far as where it is. The decaying principle of the socket will produce the disposition to fill up at the bottom whereby the stump will be pushed out; but perhaps, not till it has given some fits of the Tooth-ache. However, this circumstance does not always become a cause of the Tooth-ache.

TRANSPLANTING TEETH.

Although this operation is in itself a matter of no difficulty, yet, upon the whole, it is one of the nicest of all operations, and requires more chirurgical and physiological knowledge than any that comes under the care of the dentist. There are certain cautions necessary to be observed, especially if it be a living Tooth which is to be transplanted; because in that case it is meant to retain its life, and we have no great variety of choice. Much likewise depends upon the patient: he should apply early, and give the dentist all the time he thinks necessary to get a sufficient number of Teeth that appear to be of a proper size, &c. Likewise he must not be impatient to get out of his hands before it is advisable.

The *incisores*, *cuspidati*, and *bicuspides*, can alone be changed, because they have single fangs. The success is greatest in the *incisores* and *cuspidati* than the *bicuspides*; these last having frequently the ends of their fangs forked, from which circumstance the operation will become less perfect.

It is hardly possible to transplant the grinders, as the chance of fitting the sockets of them is very small. When indeed a grinder is extracted, and the socket sound and perfect, the dentist may, perhaps, be able to fit it by a dead Tooth.

OF THE STATE OF THE GUMS AND SOCKETS.

The first object of attention is the Sockets and Gums of the person who is to have the fresh Tooth. If the Tooth, which is to be removed, be not wholly diseased, there is great probability that the Socket will be as sound and complete as ever; but if the body of the Tooth has been destroyed some time, and the fang has been in the state of what is commonly called a stump, it has probably begun to decay on its outer surface and point; in which case the Socket will be filled up in the same proportion; if so, there is no possibility of success. But as in the operation of transplanting, the diseased Tooth is to be first drawn, it will show the state of the Socket; and the Scion* Tooth is to be left or drawn, according to the appearance on the diseased one.

If the appearance be not favourable, and it therefore be not probable that the Scion Tooth can be introduced, so as to unite in the place of a stump, I would recommend to every dentist to have some dead Teeth at hand that he may have a chance to fit the Socket. I have known these sometimes last for years, especially, when well supported by the neighbouring Teeth. Indeed this very practice is recommended by some dentists in preference to the other. But even this should not be attempted, unless the Socket is sound and pretty large, as the Tooth can otherwise have but very little hold.

Whenever there are Gum Boils, I would not recommend transplanting, as there is always in such cases a diseased Socket, although the disease has originated in the Tooth. In one or two instances, indeed, which I have seen, the Boil has been cured by such an operation.

---

* As the transplanting of Teeth is very similar to the ingrafting of trees, I thought that term might be transferred from gardening to surgery, finding no other word so expressive of the thing.

If the Gums are diseased, and become spongy, as has been described, it will be very improper to transplant, as there will be but little chance of success; also, if the Sockets have a disposition to waste; and the Tooth becomes in some degree loose; in short, the Sockets and Gums should be perfectly sound. No person should have a Tooth transplanted, while taking mercury, even although the Gums are not affected by it at the time for they may become affected by that medicine before the Tooth is fixed. I would carry this still farther; no one should have a Tooth transplanted, who has any complaint that may subject him to the taking of mercury before the Tooth is well fixed. For this reason, those who have Teeth transplanted, ought particularly to avoid for some time the chance of contracting any complaint, for the cure of which mercury may be necessary.

I would not recommend transplanting, even where mercury has been taken lately. How soon mercury may be taken after a Tooth has been transplanted, is not easily ascertained. I have known it fail from this cause, (as it seemed) after six weeks, where there was every reason to suppose that it might have been attended with success.

OF THE AGE OF THE PERSON WHO IS TO HAVE THE SCION TOOTH.

The Socket should be of its full size, and one or two grinders on each side of each jaw should be full grown, to keep the two jaws at a proper distance, which will allow the transplanted Tooth to be undisturbed by the motion of the jaw while fastening. This will be at the age of eighteen or twenty years.

It sometimes, however, happens, that a fore Tooth decays before this age, and even before it is completely formed; and therefore all the above mentioned advantages cannot be had. In such cases, it is not very material whether transplanting is practised or not, as simply to draw the diseased Tooth, will in most cases be sufficient; for the two neighbouring Teeth may be brought together, so as to fill up the space, the others following in a less degree, as has been already observed upon irregularities of the Teeth.

OF THE SCION TOOTH.

The Scion Tooth, or that which is to be transplanted should be a full grown young Tooth; young because the principle of life and union is much stronger in such than in old ones.

It will be scarcely necessary to observe, that the new Teeth should always be perfectly sound, and taken from a mouth which has the appearance of that of a person sound and healthy; not that I believe it possible to transplant an infection of any kind from the circulating juices; although we know from experience that it may be done by a matter secreted from them. The Scion Tooth should be less than what the Tooth was, the place of which it is to supply. This cannot at first be known with certainty, but it may in most cases be nearly ascertained; and that is by judging from the sizes of the bodies of the two Teeth; but as the fangs do not always bear an exact proportion to the body, it sometimes happens that this method fails. Also it is not always in our power to judge after this manner; for in some cases the body of the Tooth of the person who is to have one transplanted, shall be quite destroyed, the fang only remaining: in these cases we must judge from its correspondent on the opposite side; but even that tooth is sometimes destroyed.

It has been supposed, that we run no risk by taking the Scion Tooth from a young subject; but this is no security, for a complete Tooth is of the same size in the young as the old.*

---

* Vide Natural History, page 146, on the Growth of Teeth.

To remedy this inconvenience as much as possible, the Scion Tooth should be that of a female, for female Teeth are in general smaller than those of men; but the inconvenience still remains, whenever a female is the subject of this operation. Some women have such small Teeth that it is almost impossible to fit them. When the fang of the Scion Tooth is larger than that which it is intended to supply, it must be made smaller, and only in that part where it exceeds. But the necessity of this should be avoided, if possible; for a Tooth that is filed has lost all those inequalities which allow it to be held much faster. If, however, some part must be removed, it should be done so as to imitate the old Tooth as much as possible. The best remedy is to have several people ready, whose Teeth in appearance are fit; for if the first will not answer, the second may. I am persuaded this operation has failed from a Tooth being forced in too tight; for let us reflect what must be the consequence of such practice. A part of the soft covering of the Tooth, or lining of the Socket, is squeezed between two hard bones, so that all circulation of juices is prevented; a mortification in that part takes place; and in consequence of that a Gum Boil, and the loss of all union between Tooth and Socket; so that the Tooth drops out.

It will be hardly necessary to mention, that the sooner the Scion Tooth is put into its place the better, as delay will perpetually lessen the power upon which the union of the two parts depends.*

OF REPLACING A SOUND TOOTH, WHEN DRAWN BY MISTAKE.

It sometimes happens, that a Tooth is drawn on an idea

---

* Vide Natural History of Teeth, pages 156, 157, for an explanation of the principle upon which the success of this operation depends.

that it is diseased, because it gives pain, but appears after the extraction to be perfectly sound. In such a case I would recommend the replacing it; that there may be no loss by the operation; and the seat of the brain will probably be removed to the next Tooth. A Tooth beat out by violence, should be replaced in the same manner. This ought to be done as soon as possible; however, I would even recommend the experiment twenty-four hours after the accident, or as long as the Socket will receive the Tooth which may be for some days.

If the Tooth be replaced at any time before its life is destroyed, it will re-unite with the cavity of the Socket, and be as fast as ever.

No Tooth is excepted from this practice; for although in the Grinders there are more fangs than one, yet these fangs will as readily go into their respective Sockets as one fang would; and most probably when the Tooth has been beat out, the Sockets are enlarged by their giving way.

However, the Grinders are not so subject to such accidents as the fore Teeth, both from their situation, and from their firmness in the Sockets.

Where a Tooth has been only loosened, or shoved out in part, the patient must not hesitate, but replace it immediately. As a proof of the success to be expected from replacing Teeth, I will relate the following case.

A gentleman had his first *bicuspis* knocked out, and the second loosened. The first was driven quite into his mouth, and he spit it out upon the ground; but immediately picked it up and put it into his pocket. Some hours afterwards he called upon me, mentioned the accident, and shewed me the Tooth. Upon examining his mouth, I found the second *bicuspis* very loose, but pretty much in its place. The Tooth, which had been knocked out, was not quite dry, but very dirty, having dropped on the ground, and having been some time in his pocket. I immediately put it into warm water, let it stay there to soften, washed it as clean as possible, and then replaced it, first having introduced a probe into the Socket to break down the coagulated blood which filled it. I then tied these two Teeth to the first grinder, and the *cuspidatus* with silk, which was kept on some days, and then removed. After a month they were as fast as any Teeth in the head; and, if it were not for the remembrance of the circumstances above related, the gentleman would not be sensible that his Teeth had met with any accident. Four years have now passed since it happened. (*e*)

(*e*) [Several cases are recorded where a sound tooth extracted accidentally has been replaced and has afterwards continued in a healthy state. The success of such an operation will materially depend upon the age and constitution of the patient as shown by the history of the following cases:—

"The two following instances of the replacement of a sound temporary tooth in the lower jaw that had been accidentally but unavoidably removed in extracting a fang level with the gum and overlapped by the neighbouring bicuspid tooth well illustrates what has been said above of the effects of constitution in modifying disease:—

"Both patients were females, between fifteen and sixteen years of age, and in both cases it was a lower bicuspid which was returned to its socket. The one girl was of a bright, clear, somewhat florid complexion, having all the appearance of perfect health, and resided in London; the other lived in the country, but the countenance was pallid, the skin delicate and transparent, the stature below the average, and although she was free from any special disease, her general appearance was anything but healthy.

"In the first case the extracted tooth was rinsed in warm water, it having fallen on the floor, and immediately replaced in its socket, the patient being directed to call the next day. Accidental circumstances prevented my seeing this patient for more than three months, when one morning she brought a younger sister to the Dispensary. Upon enquiring after the tooth that had been replaced, she informed me it was a little tender for a day or two but that soon went off, and now she knew no difference between it and any of her other teeth, nor could I myself detect any alteration in its appearance any more than if it had never been removed from its socket, thrown upon the sandy floor, washed, and replaced in the mouth.

"In the second case the tooth was replaced, and when the patient was seen on the following day the parts were found to be swollen and

OF TRANSPLANTING A DEAD TOOTH.

The insertion of a dead Tooth has been recommended, and I have known them continue for many years. If this always succeeded as well as the living, I would give it the preference, because we are much more certain of matching them, as a much greater variety of dead Teeth can be procured than of living ones. But they do not always retain their colour, but are susceptible of stain. However, I have known them last for years without any alteration; and some have appeared rather to acquire a transparency, which dead Teeth in general have not.

OF THE IMMEDIATE FASTENING OF A TRANSPLANTED TOOTH.

When a Tooth has been transplanted, the next thing to be done is to fix it in that position in which it is intended to remain; that is in general to the two neighbouring Teeth, by means of silk or sea weed. If it is an *incisor* or *cuspidatus*, the silk should first be tied to the neck of one of the neighbouring Teeth, as near the gum as possible; then the two ends of the silk should be brought round upon the body of the Scion Tooth, but not so near the gum as in the former, and tied there; then it should be brought round the neck of the other neighbouring Tooth, as near the gum as possible, as in the first, and tied there. The reason of the difference of the heights of the silk recommended, must appear evident, it

painful, the tooth tender to the touch, and it was quite evident if the tooth was allowed to remain violent inflammation would supervene. There was, therefore, no alternative but to remove the tooth when the symptoms speedily subsided. In this case, if the tooth had been allowed to remain, it would certainly have produced an acute abscess, and as in the case of the little boy before mentioned, it would very probably have involved the death of a portion of the jaw bone."] (1)

(1) *The Teeth in Health and Disease.* By R. T. Hulme, M.R.C.S., F.L.S. p. 52. London, 1864.

being our intention to keep the Tooth close to the bottom of the Socket.

If the transplanted Tooth be a *bicuspis*, the same mode of tying may be followed; but the silk may be brought over its grinding surface between the two points, by which it will be better confined than in any other way. It sometimes happens, that the body of the Scion Tooth is either too long, too thick, or in such a position as to be pressed upon by the Teeth of the opposite jaw. Great care should be taken to prevent this, as the opposite Teeth constantly oppose the fastening of those which are transplanted, in every motion of the jaw. To remedy this inconvenience, we have recommended smaller Teeth than those lost; but even when they are of a proper size in other respects, they shall in some cases still touch the opposite Teeth. When this arises from length of the Tooth, a small portion may be filed off from the cutting edge with great safety. If it is owing to the thickness of the Scion Tooth, and in the upper jaw, some part may be filed off the hollow or concave surface of the Tooth, where the opposite touch. When it is owing to the position of the Teeth, the same thing may be done with propriety. By attending to this circumstance in the tying, this inconvenience may in many cases be prevented; however, if it should not be in the power of the dentist to prevent it by the above mentioned method, than he should bring them forwards by tying them to a silver plate, a little more bent than the circle of the Teeth, and resting at each end upon the neighbouring Teeth.

Where a Tooth does not exactly fit, but is too short, then there arises a difficulty with the patient whether he ought to consult propriety or beauty. The Tooth should be as much in the Socket as it can be with ease; for, although in that case it is too short, appearances must give way.

The patient must now finish the rest. He must be particularly attentive at first, and give it as little motion as possible. In many cases a soreness will continue some days, and the gums will swell; in others there will neither be soreness nor swelling.

The patient must take great care not to catch cold, or expose himself to any of the other common causes of fever; for such accidents are very likely to prevent the success of this operation. This caution is more necessary in the winter, than the summer.

The Tooth in some will begin to be fast in a few days, and the gum will cling close to it; while, in others, many weeks will pass before this happens; though the Tooth may become fixed at last.

I have seen the transplanted Tooth come a little way out of the Socket; and, without any art being used, retire into it as far as at first. The silk is to be removed sooner or later, according as the Tooth is more or less fast; in some people after a fortnight, in others not till some months after the operation.

This operation, like all others, is not attended with certain success. It sometimes happens that the two parts do not unite; and in such cases the Tooth often acts as an extraneous body*, and instead of fastening, the Tooth becomes looser and looser; the gum swells, and a considerable inflammation is kept up, often terminating in a Gum-Boil. In some cases, where it is also not attended with success, there are not these symptoms; the parts appear pretty sound, only the Teeth do not fasten, and sometimes drop out.

It also happens, that transplanted Teeth have a very singular operation performed on them while in the Socket; the living Socket and Gum finding this body kept in by force, so that they cannot push it out, set about another mode of getting rid of it, by eating away the fang till the

---

* I say often, because I do not suppose that it always acts as an extraneous body; because we know that dead Teeth have stood for years, without affecting the Sockets or Gums in the least. We may therefore suppose, that it is sometimes the case with transplanted living Teeth.

whole is destroyed, exactly similar to the wasting of the fangs of the temporary Teeth in the young subject.*

I have all along supposed, that where this practice is attended with success, there is a living union between the Tooth and Socket, and that they receive their future nourishment from this new master. My reasons for supposing it were founded on experiments on other parts,† in animals and also observations made on the practice itself; for first I observed that they kept their colour, which is very different from that of a dead Tooth; for a living Tooth has a degree of transparency, while a dead one is of an opaque chalky white.

Secondly, there are instances of their becoming diseased, in the same manner as an original living Tooth; at least the following case favours strongly this opinion.

In October, 1772, a gentleman, of the city of London, had a Tooth transplanted, which was perfectly sound, and fixed in its new Socket extremely well; about a year and a half after two spots were observed on the fore part of the body of the Tooth, which threatened a decay; they were exactly similar to specks, or the first appearance of decay, which come upon natural living Teeth. Pain is also sometimes felt in the transplanted Tooth.

But what puts it beyond a doubt is, that a living Tooth, when transplanted into some living part of an animal, will retain its life; and the vessels of the animal shall communicate with the Tooth; as is shown by the following experiments.

I took a sound Tooth from a person's head; then made a pretty deep wound with a lancet into the thick part of a cock's comb; and pressed the fang of the Tooth into this wound, and fastened it with threads passed through other parts of the comb. The cock was killed some months after; and I in-

---

* Vide Natural History, page 140.
† Vide Natural History, page 156.

jected the head with a very minute injection: the comb was then taken off, and put into a weak acid, and the Tooth being softened by this means, I slit the comb and tooth into two halves, in the long direction of the Tooth. I found the vessels of the Tooth well injected, and also observed that the external surface of the Tooth adhered everywhere to the comb by vessels similar to the union of a Tooth with the Gum and Sockets.* (*f*)

---

* I may here just remark that this experiment is not generally attended with success. I succeeded but once out of a great number of trials.

(*f*) [It is unnecessary in the present day to enter into any discussion on the merits of an operation now, I believe, wholly discontinued. From the experiments which Hunter made on the transplanting of teeth from the jaw to other situations, and the successful result of several of them, the operation in question became a favourite one with him; and it appears to have been very frequently performed either by himself or under his directions. The frequent failures which occurred, even in the operation itself, and still more the severe results which very often succeeded its performance at different periods, have very properly induced almost all subsequent practitioners to abandon its employment. Nothing but the sanguine expectations created in an ardent mind, by the interesting results which followed his first experiments, could account for a man of so sound a judgment having followed up a practice so obviously objectionable. The experiment with which this section is closed has, however, an interest attached to it far more important than its having given rise to the temporary adoption of an objectionable operation. In the result of this experiment may be found an interesting collateral argument in favour of the organised structure of the teeth, and their actual living connexion with the body. The vessels of the tooth, we are told, were well injected, and the external surface adhered everywhere to the comb by vessels. To what purpose are those vessels formed, what object can possibly be fulfilled by the existence of a vascular pulp in the interior cavity, and a vascular periosteum covering the external surface—so obviously vascular that it was well injected from the vessels of a cock's comb into which it had been transplanted—unless they are intended to nourish the bony substance of which the tooth consists, and to form the medium of its connexion with the general system?—T. BELL.]

OF DENTITION.

Teeth, at their first formation, and for some time while growing, are completely inclosed within the Sockets and Gums,* and in their growth they act upon the inclosing parts in some degree as extraneous bodies; for while the operation of growth is going on in them, another operation is produced, which is a decay of that part of the Gum and Socket that covers the Tooth, and which becomes the cause of the very disagreeable, and even dangerous symptoms, which attend this process. As the Teeth advance in size, they are in the same proportion pressing against these Sockets or Gums, from whence inflammation and ulceration are produced.

That ulceration which takes place in Dentition, is one of the species which seldom or never produces suppuration. However in some few cases I have found the Gums ulcerated, and the body of the Tooth surrounded with matter; but I believe this seldom happens till the Tooth is near cutting the skin of the Gums.

As this is a disease of an early age, and indeed almost begins with life, its symptoms are more diffused, more general, and more uncertain at such an early period, than those of any disorder of full grown people, putting on the appearance of a great variety of maladies; but these symptoms become less various and less hazardous, as the child advances in years; so that the double Teeth of the child, and still more so the second set of Teeth, or those of the adult, are usually cut without producing much disturbance.

These symptoms are so various in different children, and often in the same child, that it is difficult to conceive them to be from the same origin; and the varieties are such as seem to be beyond our knowledge.

They produce both local and constitutional complaints, with local sympathy.

---

* Vide Natural History, page 115.

The local symptoms we may suppose to be attended with pain, which appears to be expressed by the child when he is restless, uneasy, rubs his gums, and puts everything into his mouth. There is generally inflammation, heat, and swelling of the Gums, and an increased flow of saliva.

The constitutional, or general constitutional symptoms, are fever, and universal convulsion. The fever is sometimes slight, and sometimes violent. It is very remarkable both for its sudden rise and declension; so that in the first hour of this illness the child shall be perfectly cool, and in the second flushed and burning hot, and in the third temperate again.

The partial or local consequential symptoms are the most various and complicated; for the appearance they put on is in some degree determined by the nature of the parts they affect; wherefore they imitate various diseases of the human body. These symptoms we shall describe in the order of their most frequent occurrence.

Diarrhœa, costiveness, loss of appetite, eruptions on the skin, especially on the face and scalp, cough, shortness of breath, with a kind of convulsed respiration, similar to that observable in the hooping cough, spasms of particular parts, either by intervals or continued, an increased secretion of urine, and sometimes a diminution of that secretion, a discharge of matter from the *penis*, with difficulty and pain in making water, imitating exactly a violent *gonorrhœa*.

The lymphatic glands of the neck are at this time apt to swell; and if the child has a strong tendency to the scrofula, this irritation will promote that disease.

There may be many other symptoms with which we are not at all acquainted, the patients in general not being able to express their feelings. Many of the symptoms of this disease are dangerous, namely, the constitutional ones, and also those local symptoms which attack a vital part. The fever, indeed, seldom lasts so long as to be fatal; but the convulsions, especially when universal, frequently are so. Local convulsions, if not in a vital part, although often very violent, do not kill; and when any part not vital sympathizes, the patient is generally free from danger; a security to the whole being obtained by the sufferings of a part which is of very little consequence to life.

Universal sympathy seems to be the first effect of irritation, and in general appears as such in those whose local and partial sensation, and irritability, are not yet formed; for, in such subjects, when one part is irritated, the whole sympathizes, and general convulsions ensue. But as the sensations and partial irritability begin to be formed, each part, in some degree acting for itself, acquires its own peculiarities; so that when a local disease takes place in a patient that is very young, it is capable of giving a general disposition to sympathize; but as the child advances, the power of sympathy becomes partial, there not being now in the constitution that universal consent of parts; but some one part is found which has a greater aptitude than the rest to fall in with the local irritation; therefore the whole disposition for sympathy is directed to some particular part, and it sympathizes according to its own peculiar action. This arises from the different organs acquiring more and more their own independent sensations as the child grows older; and gradually losing the power of sympathizing with one another: so that by the age of six years few parts suffer but those immediately affected, and in adults, who cut their Teeth, we almost always find the pain and other symptoms confined to the part, or only local sympathy taking place, such as a swelling of the side of the face.

But as the symptoms become more confined, the suffering part is often much more violently affected, than where it has a power of taking in the other parts. Therefore we find that in adults the pain of cutting a grinder is frequently excessive, and that the local inflammation is very considerable, and often of long continuance.* This is not the case with

---

* Vide Case the third.

children; their pain does not appear to be so very considerable, and we are certain that the local inflammation is not great; that it is confined to the very parts which suffer, and is not diffused over the face; so that in children the symptoms of sympathy are often more violent than those of the parts themselves. Though it is generally a fact, that the symptoms of Dentition in adults are confined to the parts immediately injured, it is not always or certainly so; for sometimes, as will appear from case the fourth, there will be the strongest symptoms imaginable from sympathy; which seems to be owing to a peculiar aptitude in the constitution to universal sympathy. These pains in the adult are often periodical, having their regular and fixed periods from which circumstance they are often supposed to be aguish, and the bark is administered, but without effect. Medicines for the rheumatism are likewise given, with as little success; when a Tooth will appear, and disclose the cause of the complaint; and by lancing the gums the cure often is performed, but the disease will recur if the gum happens to heal over the Tooth, which it will very readily do, if the Tooth is pretty deep. As these Teeth are generally slower in their growth than the others, and more especially those which come very late, they become the cause of many returns of the symptoms. How far children under this circumstance are subject to paroxysms of the disease, is not an easy thing to determine; but from many of their sympathetic symptoms going off and returning, it would appear that they have also their exacerbation.

OF THE CURE.

The cure of diseases arising from Dentition, from their nature, can only be temporary and local, even when it is directed to the real seat of the disease; and certainly every method of cure which is not so directed, must prove ineffectual, as it can only operate by destroying the effect. Opiates, indeed, will in some degree take off the irritation, by destroying the sensibility of the part; but surely it would be better at once to remove the cause, than to be attempting from time to time to remove or palliate the effect. When the sympathy is partial, and not in a vital part, it would be better to allow it to continue than cure it, because it may by such means become universal: for instance, if it is a diarrhœa, the best way is to allow it to go on, or at least only correct it if too violent, which is often the case. I have seen cases, where the stomach and intestines have sympathized so much, as almost to threaten death. The small quantity of nourishment that the stomach could admit of was hurried off by the intestines.

OF CUTTING THE GUMS.

As far as my experience has taught me, to cut the gum down to the Teeth appears to be the only method of cure. It acts either by taking off the tension upon the gum, arising from the growth of the Tooth, or by preventing the ulceration which must otherwise take place.

It often happens, particularly when the operation is performed early in the disease, that the gum will reunite over the Teeth; in which case the same symptoms will be produced, and they must be removed by the same method.

I have performed the operation above ten times upon the same Teeth, where the disease had recurred so often, and every time with the absolute removal of the symptoms.

It has been asserted, that to cut the gum once will be sufficient, not only to remove the present, but to prevent any future bad symptoms from the same cause. This is contradictory to experiment, and the known laws of the animal œconomy; for frequently the gum, from its thickness over the Tooth, or other causes, must necessarily heal up again, and the relapse is as unavoidable as the original disease.

A vulgar prejudice prevails against this practice, from an objection, that if the gum is lanced so early, as to admit of a re-union, the cicatrised parts will be harder than the original gum, and therefore the Teeth will find more difficulty in passing, and give more pain. But this is also contrary to facts; for we find that all parts which have been the seat either of wounds or sores, are always more ready to give way to pressure, or any other disease which attacks either the part itself or the constitution. Therefore each operation tends to make the passing of the Teeth easier. (g)

When the Teeth begin to give pain, we find them generally so far formed, as to be easily discerned through the gum.

The fore Teeth are to be observed at first, not on the edge of the gum, but on the fore part, making risings there, which appear whiter than the other parts; occasioned by pressure approaching towards the surface of the gum, and it may be observed, that the gums are broader than usual. At this period the incisions must be made pretty deep, till the Tooth be felt with the instrument, otherwise little effect will be produced by the operation: and this is the general rule with respect to the depth of the incision in all cases.

When the grinders shoot into the gum, they flatten the edge of the gum, and make it broad. These Teeth are more easily hit by the instrument than the fore Teeth.

The operation should not be done with a fine pointed instrument, such as a common lancet, because most probably the point will be broken off against the Tooth, which will make the instrument unfit for going on further, if more incisions are required.

A common lancet, with its point rounded, is a very good instrument; but an instrument, something like a fleam, would be of the most convenient shape.

There is no need of any great delicacy in the operation, the gums being very insensible parts; and to cut through

---

(g) [Upon the benefits derived from lancing the gums in certain disorders which accompany the cutting of the teeth, see Dr. Ashburner's work *On Dentition and some Coincident Disorders*. London, 1834. For the opposite views, the reader may consult *Dentition and its Derangements*. By A. Jacobi, M.D. New York, 1862.]

the whole gum down to the Teeth with certainty, when they are pretty deep, requires some force.

The gums will bleed a little, which may be of service in taking off the inflammation. I never saw a case, where the bleeding either proved inconvenient or dangerous. If it ever should be troublesome, I think there could be no great difficulty in stopping it. In general no application is necessary; the gums soon unite at the most distant part from the Tooth, if it lies deep; and if it be more superficial, the thin gum soon shrinks back over the Tooth, leaving it bare, and decays.

This cutting of the *dentes sapientiæ* is often attended with an inconvenience, which does not attend the others; and this happens, I believe, only when they come very late, viz., when the jaws have left off growing. This is the want of room in the jaws for these late Teeth; a circumstance which produces an addition to the other inconveniences arising from Dentition. When it takes place in the upper jaw, the Tooth is often obliged to grow backwards; and in such a position it sometimes presses on the interior edge of the coronoide process, in shutting the mouth, and gives great pain. When it takes place in the lower jaw, some part of the Tooth continues to lie hid under that process, and covered by the soft parts which are always liable to be squeezed between that Tooth and the corresponding Tooth in the upper jaw. To open very freely, is absolutely necessary in these cases; but even that is often not sufficient. Nothing but drawing the Tooth or Teeth, will remove the evil in many cases. (*h*)

(*h*) [When the development of the lower wisdom teeth is retarded or prevented either from the position in which it presents itself, or from want of space in the jaw, it often gives rise to more serious and dangerous symptoms than those mentioned by Hunter.

Whenever the eruption of these teeth may take place, they ought to be arranged at the back of the jaw in a line with the other teeth, but sometimes they acquire a wrong direction, and the crown projects either outwards or inwards. In the one case the tooth becomes buried in

CASES.

It would be endless to give histories of cases, exemplyfying each symptom of Dentition. I shall only relate a few

---

the cheek, producing more or less inflammation, while in the other it irritates and annoys the tongue ; in either case it is necessary that the tooth should be removed ; and in doing this great caution must be exercised, for if the tooth is broken, and the fangs remain in the mouth, they may give rise to very serious symptoms, and to great bodily suffering.

Another position in which the lower wisdom tooth presents itself is with the crown coming forwards against the back of the second grinder. When this occurs, and there is want of space at the back of the jaw, the fangs being placed in the opposite direction, instead of passing downwards, produce great pressure upon the jaw bone ; in many instances violent inflammation has arisen, leading to the formation of abscesses, and to the destruction of a portion of the jaw bone. The remedy is to extract, if possible, the wisdom tooth ; but, if this cannot be effected, then the second grinder should be removed ; and the pressure being by this means taken off, the inflammation may possibly subside, but no treatment will avail until this has been done. (1)

The two following cases are recorded in the *Medical and Surgical Journal* for 1841.

A lady, 29 years of age, sought advice on account of a painful tumour on the cheek, which had existed for several months. On examination, it was found to arise from the wisdom tooth projecting horizontally outwards, and lodging in the substance of the cheek. So soon as the mouth could be opened sufficiently wide the tooth was extracted, and the patient quickly recovered.

A gentleman, 45 years of age, had suffered from an ulcer on the side of the tongue, near its base. For this the patient had been salivated, which only made it worse. It was afterwards found that the wisdom tooth on that side projected inwards, and had occasioned the ulcer on the tongue ; the tooth was removed, and the ulcer healed in a few days.

Many persons have pain and swelling of the gum while the wisdom teeth are being erupted ; in some a simple incision of the gum, reaching down to the tooth, will give relief, or it may be necessary to remove a small portion of the gum. Unless the part is very much inflamed, this operation really causes very little pain, for naturally the gum is not

---

(1) See *Des Anomalies Dentaires*, par A. M. Forget. Paris, 1859 ; also Translations of the above in DENTAL REVIEW for 1860.

which are singular; and which, being extraordinary, will the better enforce the propriety, in all cases, of the cure I have recommended.

very sensitive, although it may become most acutely so when inflamed and ulcerated.

In cases of this kind the pain and swelling usually lasts for a week or two, and subsides when the tooth has made its way through the gum; but in other cases the inflammation increases, and terminates in the formation of abscesses, which extend along the jaw bone and open on the exterior of the face by one or more apertures, through which matter is discharged, and will continue to do so until the tooth is removed. It is very important that the patient should seek advice at the commencement of these disorders, not only because of the pain and inconvenience he suffers, but also on account of the difficulty attending their treatment at a more advanced stage, owing to the inability of the patient to open his mouth, in consequence of the increased swelling and pain in the parts around the joint and about the angle of the jaw.

These severe forms of disease more frequently come under the notice of the medical practitioner than of the dentist, and we must therefore avail ourselves of the records of medicine and of surgery to illustrate this portion of dental pathology. In a lecture upon the derangements produced by the wisdom teeth, M. Roberts refers to no less than three cases that were then in the Hotel-Dieu, in all of which the patients were suffering from the impediment which existed to the development of one of the wisdom teeth, showing that these more serious complications, which sometimes accompany the process, are not so rare as would generally be supposed.

A man, aged 32 years, who was otherwise strong and in good health, had suffered for two years from pain opposite the last molar tooth on the left side of the lower jaw. An abscess formed which opened on the side of the mouth, and afterwards by another opening near the chin. A surgeon succeeded in closing this up by an operation, but it soon reformed, and has continued to discharge pus ever since. The patient then consulted a dentist, who for six weeks made use of iodine injections, probably misled by the situation of the opening and the fact of the teeth being sound, but of course the treatment was of no avail. He then consulted another dentist, and requested him to remove the tooth; this person however made excuses, and declined to operate. Lastly, the patient presented himself to an itinerant operator, who correctly referred the symptoms to the wisdom tooth, and recommended him to have the tooth in front of it removed. This was done, but did not answer, and the man remained in the same condition as before. He then entered the hospital, when it was found that a probe, passed in at the opening near the chin, close to the back of the jaw, where it struck against a hard

Case I.—A young child was attacked with contractions of the *musculi flexores* of the fingers, and also of the toes. These contractions were so considerable as to keep her

substance which proved to be the wisdom tooth. The tooth was removed, and in three days the patient was able to leave the hospital, cured.

It is astonishing how far the matter arising from one of these teeth will occasionally burrow, instances having occurred where it has even opened as low down in the neck as the collar bone. In the last case the extraction of the second molar failed because the wisdom tooth was extensively diseased; but if the operation is performed before this has taken place, as in the following case recorded by Velpeau, then it may succeed:—

A lady, aged 22, had a dull aching pain at the angle of the jaw on the left side of the face, which soon extended to the adjoining teeth, but was distinct from tooth-ache. As the pain continued to increase in intensity for several months, it was thought to be rheumatism, and was treated as such but without any good effect; then blisters and a seton at the back of the neck, kept open for a month, were tried, and opiates given, but all to no purpose. She went and resided at a watering-place for some time, but came back to Paris nothing benefited. All this time the teeth were all good in appearance, the gums healthy, and nothing denoted the eruption of a wisdom tooth. However, upon making a section into the gum, over the wisdom tooth, it was found to be arrested in its progress in consequence of the direction it had taken, the crown having come directly forwards against the posterior surface of the second molar. The second molar was extracted, and the patient immediately released from her sufferings.

Other symptoms and disorders often arise from the development of these lower wisdom teeth, such as the enlargement of the tonsils, neuralgic pains, headaches, stuttering, epilepsy, and even insanity. Dr. Fricard, when a student, was attacked, in the summer of 1821 with pain in the throat, and in the following November with severe inflammation of the right tonsil. This was subdued for a time by antiphlogistic treatment, but the pain soon returned, and continued in spite of every means, up to the year 1823. The teeth and gums appeared to be perfectly healthy, and the surgeon was about to extirpate the tonsil when it was discovered that the wisdom tooth on the affected side was not through. The gum was now freely divided, but the portions of the divided gum inflamed, and had to be removed with the knife and caustic. The tooth was thus completely freed, and the obstinate inflammation of the tonsil soon disappeared.

Dr. Ashburner attended a young woman, 19 years of age, of light hair, fair complexion, and rather stout, who for several months had fingers and thumb constantly clenched, and so irregularly, that they appeared distorted. All the common antispasmodic medicines were given, and continued for several months, but without success.

I scarified the gums down to the Teeth, and in less than half an hour all the contractions had ceased. This, however, only gave relief for a time. The gums healed; the Teeth continued to grow, and filled up the new space acquired by the scarifications; and the same symptoms appeared a second time.

The former operation was immediately performed; and with the same success.

Case II.—A boy, about two years of age, was taken with a pain and difficulty in making water; and voided matter from the *urethra*. I suspected that by some means or other this child might possibly be affected by the venereal poison; and the suspicion naturally fell on the nurse.

These complaints sometimes abated, and would go off altogether; and then return again. It was observed at last, that they returned only upon his cutting a new Tooth, this

---

perspired profusely at night; her breath was very offensive, the bowels costive, and she shouted, moaned, and talked in her sleep. One day she fell into a fit, for which a physician ordered her to be bled, and she recovered sufficiently to attend as an out-door patient at the dispensary. Three weeks after she had another fit. It appeared she had been very odd and nervous in her manner, and often cried out from cramp in the legs, which was succeeded by her thumb being drawn in towards the palm of the hand, and her fingers being clenched upon it. The patient was in a state of tetanus, that is, the body was stretched out and the muscles were perfectly rigid. When the mouth was examined, the upper wisdom teeth were in place, but those in the lower jaw were covered with a hard cartilaginous substance; this was freely cut through with the lancet, and the young woman was relieved instantly.

M. Esquirol informed Velpeau that he had a case of mental derangement where the patient was restored to reason by a crucial division of the gum which liberated the wisdom tooth.]

happened so often, regularly and constantly, that there was no reason to doubt but that it was owing to that cause.

CASE III. — A lady, about the age of five-or-six-and-twenty years, was attacked with a violent pain in the upper jaw; which at last extended through the whole side of the face, similar to a violent Tooth-ache, from a cold; and was attended with consequent fever.

It was treated at first as a cold; but from its continuance, was afterwards supposed to be nervous.

The case was represented to me from the country; and I gave the best directions, that I could, on a representation of the symptoms.

She came to London some months after, still labouring under the same complaint. Upon examining the mouth, I observed one of the points of the *dens sapientiæ* ready to come through. I lanced the gums, and the disorder gave way immediately.

A lady, about the same age, was attacked with a violent pain in the left side of her face. It was regularly periodical; coming on at six o'clock in the evening. She took the Peruvian bark, which had no effect. She took antimonials, and Dover's powder, which also were equally ineffectual. But one of the points of the *dens sapientiæ* of the upper jaw, of the same side, appearing, showed the cause, and indicated the remedy. The gums were lanced; and the pain ceased.

FINIS.

www.ingramcontent.com/pod-product-compliance
Lightning Source LLC
Chambersburg PA
CBHW032205230426
43672CB00011B/2515